Advance Praise

In this compelling study of the largest demographic exercise in the world, Trina Vithayathil reveals how the hard-won governmental promise to include caste in the 2011 Census of India was quietly thwarted by the very agents entrusted with its implementation. Marshalling evidence from an extraordinary range of sources—from cabinet ministers and bureaucrats in New Delhi to census enumerators in Karnataka—Vithayathil illustrates the consequences of caste's embeddedness in state apparatuses. This is a book with urgent lessons for India and the world; it is crucial reading for anyone who cares about caste, democracy, and the politics of data.

Joel Lee, Williams College, United States

Since the inception of census enumeration, the debate over including caste information has been highly contentious. Much of the contention during colonial times revolved around the definition of caste and its operative aspect. In the post-colonial context, this issue has become a battleground among political parties. *Counting Caste* compellingly illustrates the critical role of bureaucracy, underscoring how bureaucrats have played a crucial role in obstructing the inclusion of caste in the census. The book offers valuable insights by mapping the production of caste-wise data from top-level bureaucrats to the data-entry operators and helps us understand the complexities surrounding caste enumeration and its implications for society.

Sumeet Mhaskar, O. P. Jindal Global University, India

Counting Caste

States periodically agree to social justice reforms in response to organized demands for change. *Counting Caste* examines how and why governments make such concessions but then fail to implement them. Vithayathil unlocks the secrets of bureaucratic deflection—a process whereby political leaders and bureaucrats stall policy changes—through an in-depth examination of a caste survey in India.

Political leaders conceded to collect caste-wise data in Census 2011 for the first time in India's post-independence history. Yet, in the year that followed, bureaucrats blocked a caste count in the census and rerouted it to an inexperienced part of the government. This book uncovers the plan to gather caste-wise data in an alternative project with a history of producing poor-quality data. The case of the failed caste count highlights how state institutions evade the documentation of caste power, the continued institutionalization of castelessness—which frames caste as a problem of the oppressed and hides caste privilege and power—and ongoing efforts at resisting caste hierarchy and Hindutva domination.

Trina Vithayathil is a sociologist and associate professor of global studies and affiliated faculty in Asian studies and women's and gender studies at Providence College.

SOUTH ASIA IN THE SOCIAL SCIENCES

South Asia has become a laboratory for devising new institutions and practices of modern social life. Forms of capitalist enterprise, providing welfare and social services, the public role of religion, the management of ethnic conflict, popular culture and mass democracy in the countries of the region have shown a marked divergence from known patterns in other parts of the world. South Asia is now being studied for its relevance to the general theoretical understanding of modernity itself.

South Asia in the Social Sciences features books that offer innovative research on contemporary South Asia. It focuses on the place of the region in the various global disciplines of the social sciences and highlights research that uses unconventional sources of information and novel research methods. While recognising that most current research is focused on the larger countries, the series attempts to showcase research on the smaller countries of the region.

Books in the series:
A Woman's Job: Making Middle Lives in New India
Asiya Islam

After the Exodus: Gender and Belonging in Bangladesh's Rohingya Refugee Camps
Farhana Afrin Rahman

The Backstage of Democracy: India's Election Campaigns and the People Who Manage Them
Amogh Dhar Sharma

Questioning Migrants: Ethnic Nationalism at the Limits of Universality in Pakistan
Tahir H. Naqvi

Syndicates and Societies: Criminal Politics in Dhaka
David Jackman

Performing Sovereign Aspirations: Tamil Insurgency and Postwar Transition in Sri Lanka
Bart Klem

Legalizing the Revolution: India and the Constitution of the Postcolony
Sandipto Dasgupta

Sovereign Atonement: Citizenship, Territory, and the State at the Bangladesh–India Border
Md Azmeary Ferdoush

Memories in the Service of the Hindu Nation: The Afterlife of the Partition of India
Pranav Kohli

Freedom in Captivity: Negotiations of Belonging along Kashmir's Frontier
Radhika Gupta

Founding Mothers of the Indian Republic: Gender Politics of the Framing of the Constitution
Achyut Chetan

An Uneasy Hegemony: Politics of State-building and Struggles for Justice in Sri Lanka
Shyamika Jayasundara-Smits

The Odds Revisited: Political Economy of the Development of Bangladesh
K. A. S. Murshid

Bureaucratic Archaeology: State, Science, and Past in Postcolonial India
Ashish Avikunthak

In Search of Home: Citizenship, Law and the Politics of the Poor
Kaveri Haritas

When Ideas Matter: Democracy and Corruption in India
Bilal A. Baloch

Colossus: The Anatomy of Delhi
Sanjoy Chakravorty and Neelanjan Sircar (eds.)

Deceptive Majority: Dalits, Hinduism, and Underground Religion
Joel Lee

Simultaneous Identities: Language, Education and the Nepali Nation
Uma Pradhan

Dynamics of Caste and Law: Dalits, Oppression and Constitutional Democracy in India
Dag-Erik Berg

Crafty Oligarchs, Savvy Voters: Democracy under Inequality in Rural Pakistan
Shandana Khan Mohmand

New Perspectives on Pakistan's Political Economy: State, Class and Social Change
Matthew McCartney and S. Akbar Zaidi (eds.)

Maoist People's War and the Revolution of Everyday Life in Nepal
Ina Zharkevich

Adivasis and the State: Subalternity and Citizenship in India's Bhil Heartland
Alf Gunvald Nilsen

South Asian Governmentalities: Michel Foucault and the Question of Postcolonial Orderings
Stephen Legg and Deana Heath (eds.)

Nationalism, Development and Ethnic Conflict in Sri Lanka
Rajesh Venugopal

Politics of the Poor: Negotiating Democracy in Contemporary India
Indrajit Roy

Development after Statism: Industrial Firms and the Political Economy of South Asia
Adnan Naseemullah

Courting the People: Public Interest Litigation in Post-Emergency India
Anuj Bhuwania

Government as Practice: Democratic Left in a Transforming India
Dwaipayan Bhattacharyya

Counting Caste

Census Politics, Bureaucratic Deflection, and Brahmanical Power in India

Trina Vithayathil

CAMBRIDGE
UNIVERSITY PRESS

CAMBRIDGE
UNIVERSITY PRESS

Shaftesbury Road, Cambridge CB2 8EA, United Kingdom

One Liberty Plaza, 20th Floor, New York, NY 10006, USA

477 Williamstown Road, Port Melbourne, VIC 3207, Australia

314–321, 3rd Floor, Plot 3, Splendor Forum, Jasola District Centre, New Delhi – 110025, India

103 Penang Road, #05–06/07, Visioncrest Commercial, Singapore 238467

Cambridge University Press is part of Cambridge University Press & Assessment, a department of the University of Cambridge.

We share the University's mission to contribute to society through the pursuit of education, learning and research at the highest international levels of excellence.

www.cambridge.org

Information on this title: www.cambridge.org/9781009414111

First published 2025

Printed in India by Avantika Printers Pvt. Ltd.

Cover image: A Madhubani print by Malvika Raj depicting Dr B. R. Ambedkar's life and journey

A catalogue record for this publication is available from the British Library

ISBN 978-1-009-41411-1 Hardback

Contents

Tables and Figures

Tables

Figures

Preface and Acknowledgments

This book is born out of privilege. As a US-born member of the South Asian diaspora, my identity as a Syrian Christian has allowed me to benefit from inherited caste- and class-privilege. Syrian Christians are a "privileged minority" that falls within the roughly 8 percent of the population of India, which the 1980 Mandal Commission report identifies as "'forward' non-Hindu castes and communities."[1] My family origins in Kerala and upbringing in the United States have provided me with innumerable resources and sources of advantage.

In the lengthy period that I have taken to write this book, I have spent considerable time trying to make sense of my observational and interview data. I am indebted to the vast body of scholarship and related activism of towering figures such as Jotirao and Savitribai Phule, Iyothee Thass, and B. R. Ambedkar and continuing in more recent writings of historians, social scientists, and activists of the late twentieth and early twenty-first centuries (detailed in my bibliography) that document the nature of caste hierarchy and its durability despite ongoing change and challenge. This scholarship has allowed me to contextualize my findings, despite my many blind spots, and pushed me to ask and answer questions about the institutional structures and ideologies that help to concentrate power and perpetuate gross inequalities and indignities across generations.

I am immensely grateful to the numerous people who shared their time and knowledge with me while I was conducting fieldwork in Bengaluru and other locations throughout India between 2011 and 2016. This project is only possible because of their generosity and patience. I am also extremely beholden to research assistance from Shalini Jamuna Kotresh, Arasi Arivu, and Devanshu Singh, who created a database of newspaper coverage, while Shalini also interviewed enumerated households. I am very appreciative for a research affiliation at the Institute for Social and Economic Change (ISEC) in Bangalore that provided me with an academic home while conducting fieldwork between 2012 and 2013.

Scholarly training and networks of research and writing support have been crucial to completing this project. I studied geology as an undergraduate, but I took a course in Indian history from Eleanor Zelliot and courses from political philosopher Barbara Allen that planted numerous seeds about nationalism, caste, race, religion, gender, class, ideology, and social movements that continue to bear fruit. They, like my geology professors and classmates at Carleton, challenged me to think critically, write precisely, and make sense of the world in a historically grounded way. Faculty, staff, and graduate students at Brown University where I completed my Ph.D. in sociology, were foundational support for this project. Leah VanWey allowed me to follow my research interests wherever they took me, while Patrick Heller and José Itzigsohn formed the rest of my extremely supportive Ph.D. committee. I am also thankful for questions from my two readers—Paget Henry and Gregory Elliot. My inadequate answer to one of Paget Henry's questions was a personal reminder of the considerable work ahead to write this book. I also had several graduate student colleagues and other faculty members at Brown who served as teachers and mentors in areas foundational to this research: Michael Rodríguez-Muñiz set up a research group about knowledge and state power that created a vibrant graduate student community for this project; Ashutosh Varshney taught a course on *Indian Politics* that I audited while preparing for my preliminary exams; a conference in spring 2013—co-organized by Lina Fruzzetti, Vazira Fazila-Yacoobali Zamindar, and Bhrigupati Singh, and then graduate students Poulomi Chakrabarti, Bhawani Buswala, Jamie McPike, Gayatri Singh, and me—created an opportunity for feedback on an earlier version of chapter five from Gopal Guru, Yogendra Yadav, and Nivedita Menon, among many other conference participants; and Michael White—who was a very supportive MA thesis reader and aided my entry into graduate school—taught a research methods course in which I used census data to study patterns of residential segregation in Indian cities with Gayatri Singh (and later Kanhu Charan Pradhan). I also had a research affiliation between 2012 and 2013 at the STS Program directed by Sheila Jasanoff at Harvard Kennedy School of Government, where I previously completed a Master in Public Policy.

I am extremely grateful for writing support from Orly Clergé and Zophia Edwards who have provided invaluable feedback on every chapter of this book. Completing this project would not have been possible without their companionship, advice, and intellectual community. Amber Riaz provided excellent feedback as a professional editor on an earlier version of the manuscript. Comments from Will Toner and Ashok Danavath were very helpful in finalizing the introductory chapter. Along with the individuals already mentioned, I'm grateful to Weeam Hammoudeh, Cedric de Leon, Kara Cebulko, Diana Graizbord, Aisalkyn Botoeva, Maya Mesola, Erica Mullen, Mim Plavin-Masterman, Alissa Cordner, Mujun Zhou, Karida Brown, Kelley Smith, Tina Park, Sinem Adar, Ben Onyango, Myungji Yang, Marcelo Bohrt, Peter Klein, Stephanie Savell, Amy Teller, Yashas Vaidya, Silvia Dominguez, Kalaiyarasan A., Bindu Panikkar, and Ellen Block for feedback and community as I worked through key ideas and concepts in my dissertation, article manuscripts, and draft chapters. I have found ongoing writing support in three communities at Providence College (PC)—a faculty writing group hosted at the Writing Center by Will Toner, writing retreats hosted through the Office of Sponsored Projects and Research Compliance (SPaRC) organized by Dalila Alves and Kris Monahan, and the Women's and Gender Studies summer writing retreat. I am thankful to my initial mentors at PC, Julia Jordan Zachary and Cedric de Leon, along with program support by Chandelle Wilson, and my current colleagues at PC, Mónica Simal, Kendra Brewster, Anthony Rodriguez, Jessica Mulligan, and Eric Hirsch, along with program support by Emily Meehan, for their steadfast encouragement of my research and professional development. I am grateful for librarians at PC whose work has been instrumental to secure the sources required to complete this project and my colleagues in Global Studies, Asian Studies, Women's and Gender Studies, Sociology and Anthropology, SPaRC, the School of Arts and Sciences, and Academic Affairs who have supported my research endeavors. Student research assistants helped to collect primary and secondary documents for this project: Molly Andrus, Kellen Buckley, Christina Charie, Yingting (Jenny) Chen, Savannah Dhar, Kathryn Doner, Nicole Durant, Grace Feisthamel, Dylan Flaherty, Briana Gittens, Tatianna Medina, Lela Miller, Diana Mora, Dennis Mueñtes, Alicia Terrero, Sydaya Tompkins, and Jennifer Vargas. Lisa McNamara provided invaluable research assistance when I was completing the book proposal, while three students gave incredible insights on the full manuscript: Angie Pierre, Natasha Allen, and Ana Martinez Vargas.

I received generous funding from several sources that supported my dissertation research: a National Science Foundation Doctoral Dissertation Research Improvement Grant (DDRIG), a Fulbright Hays Fellowship, and travel grants from Brown University's Graduate School, Population Studies and Training Center (PSTC), and Graduate Program in Development. In addition, generous fellowship support throughout my five and a half years in graduate school gave me the flexibility to regularly travel to India to conduct research: an NSF Graduate Research Fellowship for six semesters; three semesters from the Brown University Graduate School; and two semesters as an NIH pre-doctoral trainee in demography at PSTC. After joining the faculty at PC, grants from

PC's Committee on Aid to Faculty Research and the School of Arts and Sciences' Summer Scholars Program provided crucial support for additional research trips in the summers of 2015 and 2016, writing during the summer of 2018, and manuscript editing between 2023 and 2024. Thank you to the Department of Global Studies at PC for support for the cover art.

Thank you to colleagues and friends for inviting me to present this research. I have given talks and shared draft chapters at the All India OBC Students Association's National Seminar on Caste Census at the University of Hyderabad, the Department of Sociology at the University of California Davis, the Political Sociology and Global South working group at the University of California Los Angeles, the Centre for Modern Indian Studies at the University of Göttingen, the Department of Sociology at the University of Georgia Athens, the American Sociological Association Annual Meeting, the Society for the Social Studies of Science Annual Meeting, the Eastern Sociological Society Annual Meeting, PC's Summer Scholars Presentations, and Brown University's conference on Questioning Marginality: Dalits and Muslims in Urban India. I am very appreciative for feedback and comments during these presentations.

Thank you to the amazing editorial team at Cambridge University Press—Anwesha Rana, Qudsiya Ahmed, Aniruddha De, and Saniya Puri—for their support, vision, thoughtfulness, and feedback throughout the process of finishing this manuscript and producing the book. Two reviewers provided very helpful comments that have significantly improved the book. I am also grateful to the series editors of *South Asia in the Social Sciences* for publishing the book. While I am immensely appreciative of all the research, writing, and editorial support that I have received for this project, all errors, omissions, and oversights in the book remain my own.

During the production of this book, I connected with the incredible Madhubani artist Malvika Raj. Her work is truly an inspiration, and I am honored and humbled that her piece "Dr. B. R. Ambedkar" is on the cover.

I am eternally thankful for my wonderful family who gave me the time, encouragement, love, and steadfast support to complete this project and made life so enjoyable along the way. Amma and Appa have always supported my path and spent lengthy periods of time in Bangalore and with our wonderful family in Kerala, so that I could focus on my research with a toddler in tow. A special thanks to my community in Bangalore, who also made it possible for me to complete research with little Markose as a companion: Sandhya Aunty, Sudhir Uncle, Rita, Seema, Matashree, Arjun, Sr. Sharlene, Joe, Sita, Rachana, and their respective families. My sister, Metty, and her family Chris, Vik, Lela, and Arjay are a constant reminder of how to create a home where the spirit, head, heart, and body are fully engaged. My life companion, Chris, and our children, Markose and

Mathai, have been endlessly supportive and patient during the many years that I have spent researching and writing this book. I am so grateful for your company, love, and humor, which fill everyday with life and meaning.

Note

1. Sonja Thomas, *Privileged Minorities: Syrian Christianity, Gender, and Minority Rights in Postcolonial India*, Global South Asia (Seattle: University of Washington Press, 2018); Backward Classes Commission (B. P. Mandal, Chair), *Report of the Backward Classes Commission* (Delhi: Government of India, 1980).

Abbreviations

AFSJ	Advocates Forum for Social Justice
BAMCEF	All India Backward and Minority Communities Employees Federation
BCC	Backward Classes Commission
BEL	Bharat Electronics Limited
BJP	Bharatiya Janata Party
BPL	below poverty line
Congress	Indian National Congress Party
CPSU	Consortium of Public Sector Undertakings
DEO	data entry operator
EB	enumeration block
ECIL	Electronic Corporation of India Limited
IAS	Indian Administrative Service
IFS	Indian Forest Service
IPS	Indian Police Service
ITI	Indian Telephone Industries
JD-U	Janata Dal United
MLA	Member of Legislative Assembly
MP	Member of Parliament
MoHUPA	Ministry of Housing and Urban Poverty Alleviation (2006–2017)
MoRD	Ministry of Rural Development
NIC	National Informatics Centre
NPR	National Population Register
OBC	Other Backward Classes
ORGI	Office of the Registrar General and Census Commissioner, India
PMK	Pattali Makkal Katchi

PSU	Public Sector Undertaking
RFQ	Request for Quotation
RSS	Rashtriya Swayamsevak Sangh
SC	Scheduled Caste
SEC survey	Socio-Economic Caste Census/survey
SRD	special recruitment drive
ST	Scheduled Tribe
UT	union territory

Prologue

In October 2023, the Government of Bihar in northern India published the preliminary findings from its first state-wide caste survey and released a full report a month later. The findings from the survey led to immediate revisions in state-level reservation policy. At first glance, the recent case of Bihar offers an alternative, and more positive, ending to this book—which follows the case of a failed nationwide enumeration of caste in India. I argue in this book that the executive bureaucracy protected the invisibility of caste privilege when it blocked the inclusion of a caste-wise enumeration in Census 2011—a process I call bureaucratic deflection. The recent survey in Bihar actually supports this book's argument since one of the executive bureaucracy's most successful strategies to prevent a caste-wise enumeration in the decennial census has been to decentralize the project of enumerating caste to state governments. Bihar's political leadership undertook the survey after the central government failed to publish caste-wise data.

Building off the success of Bihar's caste survey, the opposition Congress party seeks to prevent the third successive term of the Bharatiya Janata Party (BJP) in office at the center with a national policy platform that includes a caste census during the 2024 parliamentary election. On the one hand, it feels disingenuous that Congress has chosen to rally around this issue—given its long history of excluding a caste-wise enumeration in the decennial censuses of independent India. The Congress leadership had every opportunity to collect caste-wise data in the Censuses of 1951, 1961, 1971, 1991, and 2011. In fact, this book details how the Congress leadership conceded to do so in the lead up to Census 2011 only to backtrack on its promise and push the project into a survey with a long history of producing poor quality data. The executive leadership—which switched from Congress to BJP in 2014—never published the caste-wise data collected as part of a revamped below poverty line (BPL) survey. Both dominant political parties have refused to collect caste-wise data once in power at the center and have supported the senior bureaucracy's seeming disdain for a caste-wise enumeration in the decennial census. In this policy area, they share a common history of using the promise of a caste census to secure votes but never implementing one. This book examines the case of the failed nationwide caste count to trace the embeddedness of Brahmanism—the ideological structure of caste—within the state, albeit in forms that appear more consistent with democratic norms and values.

The case of Bihar differs from this book's story of bureaucratic deflection—as the main resistance to Bihar's caste-wise survey came from outside the government bureaucracy. Yet it offers a powerful complement, as Bihar's survey took place because the central government failed to collect caste-wise data in the census and refused to publish the caste-wise data from the BPL survey. The circulation of data from Bihar has helped to make visible caste-based power and provide direction for the implementation of social justice policies. As such, Bihar offers a roadmap for a future caste-wise enumeration in the decennial census. It reminds us of the importance of finalizing caste lists for all administrative categories prior to data collection, including the "general" administrative category (that is, privileged castes); distinguishing between groups who have benefited from reservation benefits (as well as historic non-Brahman reservations) and oppressed castes who remain excluded from state benefits and redistributive programs despite histories of exclusion; and publishing the data in a timely manner to support policies and efforts to destroy caste privilege, untouchability, and caste hierarchy—and interrelated systems of domination—all with the ultimate goal of annihilating caste.

1

Introduction

The Political Maneuvers of Caste Denial

"We were in disbelief that we won," explained a caste census activist.[1] After months of organizing, the campaign to include a caste-wise enumeration in Census 2011 had culminated in the Lok Sabha, or the lower house of Indian parliament, in early May 2010. The debate spanned three afternoon sessions in which nearly every speaker across the political spectrum supported the collection of caste-wise data in the upcoming decennial census. The leadership of the Indian National Congress Party (hereafter Congress), in power at the time through the coalition United Progressive Alliance (UPA) government, found itself backed into a corner and reversed its long-standing opposition to a caste-wise enumeration in the census. A remark by prime minister Manmohan Singh at the end of the debate, followed by finance minister Pranab Mukherjee's comment to reporters the next day that "caste will be included in the present census," publicly confirmed that the UPA government would collect caste-wise data in the census for the first time since independence.[2] The finance minister reiterated the government's position during a trip to Chhindwara the following day when he told reporters, "[T]he caste-based census was last conducted in the year 1931 and the practice should have continued in post-independence period also but it did not happen. Now the UPA government has taken an initiative in this regard."[3] Decennial censuses in independent India have limited data collection to Scheduled Castes (SCs) and Scheduled Tribes (STs)—two administrative categories consisting of Dalits, or "ex-untouchables," and indigenous communities that have faced extreme indignities, violence, and structural exclusion—to determine the size of each group's affirmative action, or reservation, quota.[4] Activists who had worked tirelessly for months to expand the caste-wise enumeration in the census were elated. Their coordinated campaign had involved public forums, conferences, speeches, rallies, and collaboration with political leaders to build a base of support at a time when both Congress and the Bharatiya Janata Party (BJP) opposed a caste count in Census 2011. Their success was tremendous given that since the 1950s, a range of commissions and organizations seeking to dismantle caste hierarchy had unsuccessfully advocated for the decennial census to collect and publish caste-wise data. Decades of effort finally met with success.

Those advocating for a caste-wise enumeration in the census desired the data for two primary reasons. First, the state required census data on Other Backward Classes (OBCs) to inform the size of the OBC reservation quota. OBCs are an administrative category that consists of diverse groups, from Dalit Muslims and Dalit Christians (that is, ex-untouchables that the state does not recognize as Dalits because of their religious identities) to Shudras (or peasant subcastes) and other groups that vary by region. A rare non-Congress-led government extended reservations to OBCs in 1990—ten years after the Second Backward Classes Commission (known colloquially as the Mandal Commission) made the recommendation and more than forty years after B. R. Ambedkar submitted his resignation as India's first law minister citing the executive bureaucracy's resistance to extend reservations to a broader group of "backward classes," among other factors.[5] As such, the central government has required caste-wise data on OBCs to implement reservations since 1992, when the Supreme Court of India upheld the 1990 central government memorandum to reserve 27 percent of central government jobs for "socially and educationally backward classes."[6] The executive bureaucracy, however, has continued to limit data collection to SCs and STs in the census despite the administrative need for data on OBCs. The government has instead estimated the size of the OBC population by using data from sample surveys and projections from 1931 census data.[7] While Mukherjee's announced change in census policy emerged in response to the activism of caste census advocates, it also squarely aligned with the administrative need for census data on OBCs following the extension of reservations two decades earlier.

Second, advocates for the caste census also argued that having knowledge about the distribution of social and economic resources by caste would support new policies and existing programs to dismantle deeply entrenched caste-based inequalities. Caste is a "brahminical institution"—a system of "domination and exclusion"—reproduced through endogamy, social distance, violence, and the absence of solidarity across groups (particularly across the line of untouchability), as well as sanctions against those who violate caste norms.[8] Caste differentiates individuals and groups hierarchically based on religiously justified beliefs and social norms, shapes peoples' lived experiences and material reality, and structures how people see themselves and relate to others. Most individuals and organizations initially pushing for the inclusion of a caste-wise enumeration in the census advocated on behalf of OBCs, but the campaign expanded to include a broader set of organizations supporting the implementation of reservation policies and the eradication of structural inequalities through a range of anti-caste policies. Censuses in independent India have lumped together by default all "non-SCs/STs"—ranging from Dalit Muslims to caste-elite Brahmans—by limiting enumeration to SCs and STs. The resulting data have rendered invisible caste-related differences in opportunities and accumulated wealth and resources across generations. The census has helped to obscure the advantages of caste elites. The census would now produce detailed caste-wise data on elites, along with OBCs and other caste-oppressed groups—on a range of indicators including literacy, highest level of education completed, occupation, household amenities, and household assets.[9] The concession to change census policy on caste made sense for the successful implementation of OBC reservation policies and to support a broader set of anti-caste efforts to dismantle intergenerational inequalities and privilege related to caste hierarchy.

Resistance to the announced policy change soon emerged in and outside the state. A few weeks after the finance minister's announcement, a group of prominent figures and private citizens started the *meri jati Hindustani* (my caste is Indian) campaign to oppose the enumeration of caste in the census. Critical views of the policy decision filled national newspapers and television news shows. Much of the public coverage involved experts from academia, policy think tanks, and other knowledge-making institutions, debating whether caste data were needed and if the collection of caste data would reinforce an outdated social identity. At the same time, editorial pages of leading English, Hindi, Marathi, Kannada, Tamil, Telugu, and Malayalam newspapers published editorials both opposing and supporting the decision to enumerate caste in Census 2011. Ongoing resistance to the announced policy change by celebrities, academics, and the public campaign in Delhi remained the focus of national media coverage.[10] The media also revealed that the union cabinet was overwhelmingly against the collection of caste-wise data in Census 2011, and that the Indian census bureau remained staunchly opposed.[11] The long-standing institutional resistance to include a full caste count in the decennial census became highly visible during the summer of 2010. Privately, political leaders began to look for an alternative path to avoid the collection of caste-wise data in the upcoming census.

Internal bureaucratic negotiations in the year that followed blocked a caste-wise enumeration in Census 2011. The Office of the Registrar General & Census Commissioner, India (ORGI), which was responsible for the design and execution of decennial censuses, had little desire to deviate from its well-worn path and advocated for the collection of caste-wise data elsewhere. A series of improbable maneuvers between June 2010 and May 2011 kept the caste-wise enumeration out of Census 2011, at which point the political leadership publicly announced that they were merging the caste "census" with the Below Poverty Line (BPL) survey.[12] The Ministry of Rural Development (MoRD) had carried out the BPL survey three times previously to identify rural households eligible for anti-poverty programs and benefits. However, the BPL survey differed from the decennial census in its purpose, scope, and lack of constitutional mandate. In addition, anti-poverty experts and activists had long criticized the BPL survey for failing to identify rural households most in need of support.[13] The final resting spot for the caste count in the revamped BPL survey became a poor replacement for the census, and the project bore little resemblance to the original demands of caste census activists. While the executive leadership renamed the BPL survey the Socio-Economic Caste (SEC) Census, I refer to the project as the SEC survey throughout this book. The SEC survey remained distinct from the decennial census, and therefore to call it a "census" obscures its mandate, legal standing, and operational home. The MoRD and the Ministry of Housing and Urban Poverty Alleviation (MoHUPA)—not the ORGI—oversaw the ground operations to collect BPL and caste-wise data from every household in the country. While data collection began in July 2011 and the publication of BPL data occurred in 2015 and 2016, caste-wise data from the project still remain unpublished.[14]

This book explores why Census 2011 did not include a caste-wise enumeration following the concession by the Congress leadership in May 2010. In addition, it seeks to answer how and why the collected caste-wise data remain unavailable more than a decade after the completion of the SEC survey. I set out to explain how a victory by caste census advocates turned into an inconsequential project that failed to result in the public dissemination of caste-wise data. Why did caste-wise data never publicly materialize?

Two bodies of literature offer partial explanations for the case of the failed caste count. Scholars of caste and politics have documented the role of caste as a central organizing identity in democratic politics, from the mobilization of vote banks to the distribution of services and resources, and its importance in electoral calculations and compromises.[15] More broadly, this body of scholarship has shown the democratizing effects of oppressed-caste social movements, the entry of oppressed castes into formal politics, and the roles that related movements and political leaders have played in challenging caste-based domination.[16] From this perspective, caste census advocates secured a public concession from Congress leaders as a compromise of electoral politics. While Congress (and the BJP) remained ideologically opposed to a caste-wise enumeration in the census, they agreed to the policy change to secure future votes and political support from OBCs. The public concession to collect caste-wise data in the census was part of a political strategy to secure the "OBC vote." The eventual status quo in census policy realigned with Congress's (and the BJP's) true position, and the party leadership wagered that OBC support would remain even if they divorced the caste count from the decennial census and inserted it into another state project. While the electoral politics of caste may help to explain the first part of the story—that is, how and why caste census activists targeted political leaders to secure a change in census policy and why political leaders conceded—it fails to explain the bureaucracy's resistance. Why did the executive bureaucracy chart an alternative path in the aftermath of a clear concession by political leaders? In addition, other parts of the state, as well as nonstate actors, supported the status quo with regard to census policy on caste. The scholarship on Indian politics tends to minimize the role of bureaucrats and other experts in actualizing the demands made by citizens and movements to dismantle systems of oppression.[17]

A second body of social science research on technocracies helps to illuminate how the increased power of experts and expertise undercuts transparency and public participation in policymaking.[18] Political sociologists and scholars of science and technology studies have traced the rise of experts to show how technocracies threaten the ability for citizen-voice to shape the creation, design, and monitoring of public policies and programs. The increasing power of technocrats in contemporary democracies helps to explain how senior administrators, particularly census bureaucrats (that is, the experts in this case), secured their wish to maintain the status quo in census policy on caste despite the demands made by activists, politicians, and other non-experts. At the same time, the expansion of affirmative action to OBCs took place almost twenty years earlier. As such, technocratic power—independent of caste—should have desired caste-wise data from the census to estimate the size of the OBC population. Instead, a powerful unit within the bureaucracy exerted considerable effort to prevent a caste count in the census. Scholarship on the rise of the technocratic state fails to explain why senior bureaucrats continued to resist a caste-wise enumeration in the census nearly two decades after the expansion of reservation policy.

Explanations that focus on the compromises of party politics or the power exercised by technocrats are important yet incomplete. Scholarship in both areas remains less attentive to how and why the executive bureaucracy would choose to block a policy that was in line with the state's administrative need for data. The scholarship on technocracies generally does not attend to how caste, race, gender, sexuality, and other identities along which power and exclusion operate are integral to historical and contemporary formation of bureaucratic expertise and how experts tasked with creating and implementing policies to dismantle structural inequalities reproduce them. Social science research, more broadly, fails to sufficiently consider how the embeddedness of caste-like systems in the state and other key institutions of society challenges the possibility for democracies to correct themselves of histories of exclusion.

To answer why caste-wise data have never materialized, I contextualize the case in a different side of the politics of caste in which the proposal to count caste in the census became a battleground to protect the invisibility of caste privilege. After a brief discussion on terminology, I turn to critiques of Brahmanism written by scholars more than a century ago who document the institutionalization of caste power within the colonial state. I then explore how the central bureaucracy in independent India has continued to deflect demands for change, while seeming to align with movements to dismantle discriminatory systems. The politics surrounding the collection of census data on caste offers a sociological view into how exclusion operates within the administrative bureaucracy and other institutions of power, in response to demands for recognition, respect, and equity by historically oppressed groups.

A Note on Terminology

Since language is power, anti-caste movements and activists have long approached the language of caste critically. Caste abolitionist activist and author Thenmozhi Soundararajan uses the language of "caste apartheid" instead of the more neutral term "caste system" to challenge the common Brahmanical view that there are valuable and useful aspects of caste. Caste apartheid references the historical connection between India and the architects of South African apartheid who looked to Brahmanism as a successful "system of despotism" during their creation of a system of racial caste.[19]

Movements against caste oppression have employed language to challenge the hierarchical nature of caste and makes visible the unequal distribution of power. The term "*lower* caste" is sometimes used to refer to Shudras (which roughly corresponds to the state administrative category of OBC), and other times it is used colloquially to refer to Dalits, Adivasis, and Shudras (that is, SCs, STs, and OBCs, respectively). Yet terms such as "*lower* castes" or "*backward* castes" describe groups of people as less than, or inferior to, others. Alternative language to collectively describe Dalits, Adivasis, and Shudras includes oppressed castes or Bahujan—which means "the majority of the people"—and also religious minorities. Anti-caste activists and organizations have similarly challenged the language of "*upper* castes" or "*forward* castes" (which roughly correspond to the state administrative "general" category), as this caste hierarchical language deems groups of people superior to others. Instead, the language of dominant castes or oppressor castes names power and structural advantage. In this book, I use such language that acknowledges and describes the relational nature of caste apartheid as a system of domination.

The Art of Concealing Caste Power

The case of the failed caste census reveals how the executive bureaucracy appears to support the implementation of projects that make caste power visible, yet ultimately subverts the completion of these projects. I introduce the concept of *bureaucratic deflection* to describe this phenomenon. I then trace the origins of bureaucratic deflection by drawing upon the writings of anti-caste scholars who documented the embeddedness of Brahmanism in political institutions during the colonial period, as well as the emergence of castelessness, which makes caste a problem of oppressed groups while obscuring caste privilege. The strengthening of ideologies of castelessness and Hindutva in independent India further supports the successes of bureaucratic deflection—or the stalling of projects that seek to make visible, critique, or destroy caste privilege.

Bureaucratic Deflection

I use the case of the failed caste count to document how bureaucratic deflection unfolds. I define *bureaucratic deflection* as the process by which the administrative bureaucracy publicly complies with the demands of historically excluded groups but privately maneuvers, avoids, stalls, or resists the implementation of policies and programs that challenge the unequal distribution of power and resources.[20] The study of how bureaucratic deflection progresses opens "the black box" of what happens following movements to bring about institutional change. Social science research has explored the intended effects of social movements and the types of organizing and movements that are more likely to bring about change, yet limited research has focused on the implementation of secured social justice reforms.[21] That is, what occurs in the aftermath of the state's concession to change policy?

Bureaucratic deflection unfolds in the less public spaces of the administrative bureaucracy. It results from internal ideological opposition to an agreed-upon change, as well as organizational indifference and inertia that perpetuate the status quo. In some instances, the implementing unit might support the policy change but lack resources to counter inertia or resistance from more powerful units. The implementing unit actively blocks or delays the project's success in other instances. Through a diversity of mechanisms that emerge relationally, the bureaucracy becomes the burial ground for programmatic efforts to dismantle structural inequalities and redistribute power and resources more equitably.

One of the origins of bureaucratic deflection is the central bureaucracy's historic dominance by Brahmans and other caste elites. In 1953, over 70 percent of senior administrators were "*forward* castes," as were most ministers and senior political leaders of the Congress Party, while these groups accounted for a small fraction of the population.[22] In the top tier of the central bureaucracy, the combined population of SCs and STs was less than half of 1 percent in 1953.[23] Caste elites accessed the best available education and secured political and administrative positions in the colonial government. As such, the key personnel change between the pre- and post-independence periods involved their takeover of the highest tier of the administration from the English. Oppressor castes now had social, administrative, and political power, as Ambedkar had forewarned, and control of the political ideology for independent India.[24] To fight their dominance of the bureaucracy, Ambedkar designed a system of reservations in independent India to ensure that excluded groups could access seats in colleges and universities and coveted government jobs.

Yet the diversification at the top of the bureaucracy has been sluggish. In 1947, the Government of India reserved 12.5 percent of posts for SCs in positions recruited by open competition (the reservation quota increased to 15 percent in 1970) and 16.67 percent of posts for SCs in jobs filled on an all-India basis other than by open competition. For STs, the Government of

India reserved 5 percent of positions in 1947 (and increased the reservation quota to 7.5 percent in 1970).[25] After the judiciary decided that caste-based reservations were illegal in 1951, Ambedkar worked with elected leaders to pass an amendment to the Constitution that allowed for the reimplementation of reservations in central and state government jobs and in higher educational institutions.[26] Legal scholar Marc Galanter argues that caste-elite administrators and clerks entrusted to implement reservation policy saw it as a very low priority and failed to develop systems to recruit and promote historically oppressed groups into the upper tiers of the bureaucracy.[27] They believed that Dalits and Adivasis had inferior abilities and resisted implementing reservations to prevent the bureaucracy from becoming "inefficient."[28] After twenty-five years of reservations, SCs had secured only 3.4 percent and STs only 0.62 percent of group A positions—Dalits and Adivasis remained concentrated in clerical or low-wage work in the bottom tiers of the bureaucracy.[29] In response to the slow change, activists pressured the government to expand reservations in promotion to support movement into the higher tiers of the bureaucracy. In 1974, the government supported reservations for SCs and STs in transitions from Group C to Group B, within Group B, and from Group B to Group A.[30] Activists also compelled the government to fill the enormous backlog of reserved positions. Special recruitment drives (SRDs) sought to address the backlog in vacancies for reserved positions. More Dalits and Adivasis entered the upper tiers of the bureaucracy following the implementation of SRDs, while the extension of reservations in promotions worked to support the movement of Dalit and Adivasi employees into higher administrative positions. Yet the culture often remained hostile. Caste-elite supervisors and colleagues have made individuals who secured government jobs through reservations feel less qualified, devalued their work, deemed them undeserving, and delayed or overlooked their promotions. In contrast, individuals who secured a government job without reservations have been viewed as "meritorious" and "deserving"—without attention to intergenerational advantages, including access to elite educational institutions, financial resources, and social networks familiar with test preparation and jobs interviews, which translate into advantages during entry and promotion into the higher tiers of the bureaucracy.

As such, the culture and composition of the central bureaucracy have been slow to change.[31] Despite nearly seventy years of central government reservations, the representation of SCs and STs in group A jobs remained below the reservation quotas in 2015 (that is, 13.3 percent for SCs and 5.9 percent for STs) (Table 1.1).[32] Similarly, recent recruitment data show that

Table 1.1 Scheduled Caste (SC) and Scheduled Tribe (ST) Employment in Central Services (1953–2015)

A. SC and ST Representation in Central Services, Reservation Quotas, and Percentage of Total Population

Year	SCs in central services (%)	Reservation quota for SCs in open competition (%)	SCs in total population (%)	STs in central services (%)	Reservation quota for STs in open competition (%)	STs in total population (%)	Source
1953	13.82	12.5	15.32 (1956)	1.26	5.0	6.23 (1956)	a, b
1965	13.17	12.5	14.64 (1961)	2.25	5.0	6.80 (1961)	b, d
1975	13.84	15.0	14.59 (1971)	2.94	7.5	6.93 (1971)	b, d
1985	16.83	15.0	15.75 (1981)	4.66	7.5	7.76 (1981)	d, f
1995	17.43	15.0	16.48 (1991)	5.78	7.5	8.08 (1991)	c, d, f
2005	16.62	15.0	16.20 (2001)	6.42	7.5	8.20 (2001)	c, d
2015	17.55	15.0	16.63 (2011)	8.37	7.5	8.61 (2011)	c, e

(Continued)

Table 1.1 (*Continued*)

B. SC and ST Representation by Tier in Central Services

Year	Group A*		Group B*		Group C*		Group D*		Source
	% SC	% ST	% SC	% ST	% SC	% ST	% SC	% ST	
1953	0.35	0.10	1.29	0.24	4.52	0.46	20.53	1.84	a
1965	1.64	0.27	2.82	0.34	8.88	1.14	17.75	3.39	d
1975	3.43	0.62	4.98	0.59	10.71	2.27	18.64	3.99	d
1985	7.30	1.73	10.03	1.57	14.87	4.20	20.80	5.7	d
1995	10.15	2.89	12.67	2.68	16.15	5.69	21.26	6.48	d
2005	11.80	4.30	13.70	4.50	16.40	6.50	18.30	6.9	d
2015	13.31	5.89	16.27	6.75	17.35	8.60	42.92*	6.05	e

Source: (a) Backward Classes Commission (K. Kalelkar, Chair), *Report of the Backward Classes Commission*, 17. (b) Galanter, *Competing Equalities*, 133. (c) National Commission for Scheduled Tribes, "NCST Handbook" (Government of India, 2015), 66. (d) Government of India Ministry of Personnel, Public Grievances and Pensions, "Brochure on Reservation for SCs, STs & OBCs in Services," 9–10. (e) Government of India Ministry of Personnel, Public Grievances and Pensions, *Annual Report 2016–2017*, 53. (f) V. S. Verma, "Census of India 1981: A Handbook of Population Statistics" (Registrar General & Census Commissioner, India, 1988), 163. (g) Amulya Ratna Nanda, "Census of India 1991: Union Primary Census Abstract for SCs and STs, Paper 1" (Registrar General & Census Commissioner, India, 1993), 2, 6.

Note: * The original "Class I–IV" classifications of the administrative tiers in were replaced by "Groups A–D" classifications in the 1990s. The broad description of each tier is as follows: Class I/Group A (senior administrative), Class II/Group B (other administrative), Class III/Group C (clerical), Class IV (attendants, peons). After a more recent change in the system of classification, the 2015 data for Group C consists of all clerical workers and attendants (except Safai Karamcharis, or sanitation workers) while the 2015 data for Group D only includes sanitation workers.

the appointment of SCs, STs, and OBCs is approximately half their respective reservation quotas in the all-India civil services—the Indian Administrative Service (IAS), the Indian Police Service (IPS), and the Indian Forest Service (IFS)—whose officers hold the top civil service positions in the central and state governments. Between 2018 and 2022, the percentage of appointments in the IAS, IPS, and IFS has been at 7.7 percent for SCs, 3.8 percent for STs, and 15.9 percent for OBCs out of reservation quotas of 15 percent, 7.5 percent, and 27 percent respectively.[33] The lack of reservations at the secretarial levels—posts which are filled by senior IAS officers—has led to an even greater paucity of SCs and STs in these influential positions. Among the IAS officers appointed to secretary-level roles in March 2011, only 9.6 percent of joint secretaries (46 out of 477), 3.8 percent of additional secretaries (4 out of 108), and 2.5 percent of secretaries (4 out of 149) were SCs or STs. Thus, at the highest tiers of the administrative bureaucracy, SC and ST officers remain underrepresented in group A positions, in the IAS and other all-India services, and in secretarial-level appointments. Seemingly caste-neutral ideologies of efficiency and meritocracy mask the historical processes of discrimination, indignities, and exclusion that maintain institutional control of the bureaucracy by caste elites. Their dominance similarly creates a hostile environment for efforts to implement a caste-wise enumeration in the decennial census.

The executive bureaucracy has been lethargic in the implementation of programs to address caste-based exclusion in higher education, which also helps to maintain caste-elite dominance of the upper tiers of the bureaucracy. While there are numerous examples to cite in this area, I highlight one program that also provides a case study of how bureaucratic deflection operates.[34] Social science scholars Ashok Danavath and Jyoti Bania trace the ineffectiveness of a government program aimed at providing overseas graduate scholarships to groups historically excluded from higher education.[35] The Government of India constituted the National Overseas Scholarship several decades ago to facilitate upward mobility of low-income students from Dalit and Adivasi communities by providing financial aid for master's, Ph.D., and postdoctoral studies in top-ranked universities around the world.[36] Despite the creation of this scholarship program to diversify higher education, the bureaucracy has deflected the implementation of the program, muting its potential impact. Danavath details the unresponsiveness of the bureaucracy, discrimination during the application process, and how the scholarship's fixed amount fails to adjust to higher living costs.[37] The National Overseas Scholarship for Scheduled Tribes administered by the Ministry of Tribal

Affairs also remains severely under-resourced but at the same time has underutilized its budget. The unused funds are particularly puzzling given that the scholarship selection committee has rejected students who have been accepted by top global universities and have secured fellowships (which would cover a portion of the expenses) and supervisors.[38] In addition, irrelevant and discriminatory questions by the selection committee during the interview process perpetuate a system where applicants already known to the interview committee have greater success in securing a government scholarship even if they have not secured admission letters, while highly qualified applicants with letters of admission are rejected.[39] The timing of the scholarship application is often misaligned with the admissions cycle for overseas graduate programs, leaving accepted students unable to attend highly competitive programs. Despite all of these barriers, Dalit and Adivasi scholars have attended top-ranked universities around the world and produced groundbreaking research, including historical research and social science scholarship that have brought global attention to systemic caste-based violence and everyday forms of resistance—continuing the legacy of Ambedkar.[40] Over the last several years, additional policy modifications have tried to limit the scope and reach of the program by increasing the required grades for eligibility and discontinuing the opportunity to study Indian culture, heritage, and history. By excluding the study of Indian history and culture, the executive bureaucracy is in essence disallowing support for research that makes caste apartheid visible. As Danavath argues, "this policy, by implication, excludes students belonging to caste-marginalized communities from critically engaging with caste-based oppression." The bureaucracy has effectively limited the scope, reach, and impact of a scholarship program intended to create educational access for groups systematically excluded from higher education, support groundbreaking scholarship, and diversify the elite composition of academia.

The embeddedness of Brahmanism within the state allows for democratically secured policy changes to become sabotaged by the bureaucracy during implementation. By simply "dragging its feet," the bureaucracy has been able to delay projects by decades or limit the scope and effectiveness of programs seeking to challenge caste-based exclusion. To understand the origins of this type of resistance, I trace how Brahmanism became embedded in political institutions during the colonial period and its ongoing presence in independent India, despite taking on new forms that appeared consistent with democratic ideals.

The Institutionalization of Brahmanism

To understand why the caste census failed, a robust conceptualization of Brahmanism is necessary. Yet, historian G. Aloysius explains, "there has been much squeamishness within academia in using Brahmanism as an explanatory category in macro level political studies."[41]

Anti-caste scholars and activists conceptualized Brahmanism as the ideological structure of caste in their nineteenth- and early-twentieth-century writings. Scholar and playwright G. P. Deshpande writes in an essay introducing Jotirao Phule's writings, "[F]or Phule, Brahmanism was historical, constructed over time, and since it was the ideology of oppression and dominance, it had to be opposed and ultimately smashed."[42] Phule, a prolific writer and activist who along with his wife, Savitribai, created schools for caste-oppressed groups and girls in mid-nineteenth-century Pune, sought to destroy the Brahmanical structure "not simply because it would free the *shudraatishudras* [that is, oppressed castes], but because it would free society as a whole."[43] Phule wrote a scathing account of Brahmanism in his 1873 text entitled *Slavery in the Civilized British Government under the Cloak of Brahmanism.* Phule documents how a Brahmanical ideology affects all:

> The highest rights, the highest privileges and gifts, and everything that would make the life of a Brahman easy, smooth going and happy—everything that would conserve or flatter their self-pride—were specially inculcated and enjoyed, whereas the Sudras and the Atisudras were regarded with supreme hatred and contempt, and the commonest right of humanity were denied to them. Their touch, nay, even their shadow, is deemed a pollution. They are considered as mere chattels, and their life of no more value than that of meanest reptile.[44]

Phule describes a system that creates fundamentally different life experiences and opportunities across castes in which the advantages and privileges of Brahmans and other caste elites are directly tied to the enslavement, inhumanity, and disrespect that oppressed castes experience—one is not possible without the other. A half century later, B. R. Ambedkar continues to document the relational nature of caste and important variations in lived experiences across groups:

> Each caste takes its pride and its consolation in the fact that in the scale of castes it is above some other caste. As an outward mark of this gradation, there is also a gradation in social and religious rights technically spoken of in *Ashta-dhikaras* and *Sanskaras*. The higher the grade of a caste, the greater the number of

> these rights; and the lower the grade, the lesser their number. Now this gradation, this scaling of castes, makes it impossible to organize a common front again the Caste System.... All are slaves of the Caste System. But all the slaves are not equal in status.[45]

Ambedkar argues that caste hierarchy mentally enslaves all and justifies a system of differential power, status, and rights across groups. This finely gradiated system creates a challenge for collective organizing against caste hierarchy. Like Phule before him, Ambedkar emphasizes how Brahmanism is not restricted to Brahmans but undergirds the entire system:

> By Brahmanism I do not mean the power, privileges, and interests of Brahmans as a community. That is not the sense in which I am using the word. By Brahmanism I mean the negation of the spirit of Liberty, Equality and Fraternity. In that sense it is rampant in all classes and is not confined to the Brahmans alone though they have been the originators of it. This Brahmanism which pervades everywhere and which regulated the thoughts and deeds of all classes is an incontrovertible fact. It is also an incontrovertible fact that Brahmanism gives certain classes a privileged position. It denies certain other classes even equality of opportunity.[46]

Ambedkar contends that Brahmanism negates the possibility of equality of opportunity and treatment. It gives groups deemed superior privileges and power while simultaneously creating subjugated and "untouchable" groups.[47]

Anti-caste activists also described and denounced the co-constitutive role of patriarchy and Brahmanism in the latter half of the nineteenth century. Tamil Buddhists, such as M. Masilamani, argued that the subjugation of women and of oppressed classes were the sources of caste-based power.[48] Jotirao Phule similarly spoke out against the system of gendered caste oppression, including the discrimination that Brahman women experienced in enforced widowhood that created "private and public prostitutes."[49] Writing a half century later, Ambedkar also theorized caste and gender as deeply "entangled" within a system of graded inequalities and named endogamy or "the absence of intermarriage," along with "the violent control of surplus woman's sexuality" and the ideologies that justify such control, as central to the origins of caste.[50] Sociologist Sharmila Rege argues, "Ambedkar saw caste's exclusionary violence and subjugation of women inherent to the very processes that lead to caste formation."[51] Historian Shailaja Paik documents how "the control of sex and female sexuality leads to the social reproduction of caste" as she connects the "big history" of Ambedkar and other Dalit leaders who "operated overtly in the public sphere" with the undocumented and therefore largely unknown lives of Dalits, such as Pavalabai, a Tamasha (traveling public theatre) artist who "within the context of caste slavery … desired to govern herself" and "continually contested caste patriarchies."[52] Despite limited historical documentation of caste-oppressed women who fought against a system of gendered caste apartheid during this period, feminist scholars such as Rege, Paik, and Uma Chakravarti have detailed how Brahmanism and its constitutive patriarchy justified determinations of worth and humanity based on gender and caste.[53]

A Brahmanical ideology also shaped the nationalist movement in India. Scholar-activists writing in the late nineteenth and early twentieth centuries traced the processes by which Brahmanism became embedded in political institutions. They highlighted how the labor of caste-oppressed groups was central to the construction of state institutions, yet these same groups were blocked from accessing "public" spaces and resources. Caste-oppressed groups also experienced disrespect and indignities in government offices and had limited means of redress given the dominance of caste elites and Brahmanical culture in the bureaucracy.[54] The control of the government bureaucracy by Brahmans and Brahmanism shaped state knowledge-making practices, including the census, and blocked access to schools for caste-oppressed groups. I discuss these points in greater detail in the next chapter to document the embeddedness of Brahmanism within the colonial state.

Anti-caste leaders continuously critiqued the mainstream nationalist movement for fighting for *swaraj* (self-rule), which attacked foreign colonialism, while maintaining and further exacerbating a system of internal caste domination. Feminist scholar and activist V. Geetha and scholar and activist S. V. Rajadurai discuss how Iyothee Thass, a Tamil Buddhist and outspoken activist and writer on behalf of people deemed "pariahs" and "outcastes" in colonial Madras, "interrogated the politics of Brahmanism and the Brahmanical tenor of nationalism" in his scathing critiques of the Congress Party.[55] Iyothee Thass referred to Congress as a party of "Brahamanas" after his unsuccessful effort to bring a petition for the removal of caste distinctions before Congress in 1891.[56] He continuously spoke out against Congress's plans for self-government, which remained riddled with caste inequalities and prejudices. In a series of columns on "*swadeshi* reform" in his weekly magazine, *Oru Paisa Tamizhan*, between 1907 and 1908, he critiqued anew the phase of Congress organizing that put forward the twin ideals of *swaraj* and *swadeshi* (locally made products). He argued that the mainstream nationalist movement's boycott of foreign goods to preserve all things Indian represented an "uncritical pride in one's self and castehood."[57] He asserted that in a society where norms of untouchability prevented caste Hindus from even helping a dying child whom they deemed untouchable, all talk of self-rule and self-reliance was insincere and benefited the power of caste elites.[58] In the emerging ideological streams within Congress—Hindu nationalism, Gandhian *swadeshi*, and Nehruvian secularism—there was little attention to liberatory identities for oppressed castes.[59] Sociologist Gail Omvedt asserts that Congress's nationalist ideologies were foundationally Brahmanical in their orientation: "its structures of dominance, both organizationally and ideologically, were such as to draw its members into a Brahmanic-Hindu framework of interpreting the 'Indian' and into incorporation within a capitalist system" which could not "provide an Indian identity that was liberatory for Dalits and low castes."[60] Shailaja Paik describes how M. K. Gandhi's paradoxical views of caste advocated to end untouchability and grant Dalits access to certain public institutions (for example, education and water tanks) but did not fundamentally attack the caste structure, since he believed caste provided the labor needs of society and social order.[61] Political scientist Christophe Jaffrelot traces how Gandhi's views on caste shaped the conservative orientation of the Congress's political platform, which opposed untouchability but failed to challenge caste hierarchy. Given the caste politics of Congress, Ambedkar discussed the challenges of forming a nation with a deeply embedded caste system during the Constituent Assembly debates that finalized the Constitution of India:

> I am of the opinion that in believing that we are a nation, we are cherishing a great delusion. How can people divided into several thousands of castes be a nation? The sooner we realize that we are not as yet a nation in the social and psychological sense of the world, the better for us.... In India, there are castes. The castes are anti-national in the first place because they bring about separation in social life. They are anti-national also because they generate jealousy and antipathy between caste and caste. But we must overcome all these difficulties if we wish to become a nation in reality. For fraternity can be a fact only when there is nation. Without fraternity, equality and liberty will be no deeper than coats of paints.[62]

For Ambedkar, the challenge of nationhood meant political institutions needed to actively work to destroy caste hierarchy, abolish untouchability and patriarchy, foster fraternity, open pathways to education and government employment and other previously excluded jobs, and ensure political representation for caste-oppressed groups. Brahmanism precluded social equality—and without social equality, political equality could not be enacted.[63] Those who controlled political institutions in decolonizing India did not adequately heed Ambedkar's call. To help explain the administrative resistance to a caste-wise enumeration in the early twenty-first century, I focus next on Brahmanism's resilience in ways that appeared aligned with democratic norms.

The Castelessness of Nationalism

Built into the nationalist project of constructing a representative democracy was the "unmarking" of caste identities of political leaders who were overwhelmingly from oppressor-caste backgrounds. Sociologist Satish Deshpande describes how, during the height of the independence movement in the 1930s, the leaders of the overwhelmingly caste-elite Congress Party muted their caste identities to become credible, *all-India* representatives in response to oppressed castes' assertions for equality, dignity, and the dismantling of caste hierarchy.[64] Once "unmarked," these caste-free elites became modern, universal subjects whose successes reinforced narratives of meritocracy and individualism while obscuring privilege and structural advantages.[65] Anthropologist Ajantha Subramanian has traced how caste privilege transforms into merit, thus reinforcing the institutionalized view of caste solely as a source of disadvantage—and not as privilege.[66] In contrast, leaders and members of oppressed-caste communities remained caste-marked in political life. As such, *castelessness* emerged as a subjectivity that began to refigure caste primarily as a "problem" of the oppressed. Deshpande argues that the institutionalization of castelessness through law and policy minimally required that caste-based privileges wear "the garb of secular modernity," as "caste was henceforth to be recognized only as sources of disadvantage or vulnerability, not as a source of privilege or advantage."[67] Feminists, critical race theorists, and anti-caste scholar-activists have shown how promises of universalism ignore differences in political projects that challenge the existing distribution of power versus programs that gloss over or reinforce hierarchies of power. Legal scholar Dorothy Roberts argues that "universalism attempts to solve the problem of racial inequality within programs that paper over, rather than uproot, the social forces that structure the current racially stratified system and that perpetuate racial injustice."[68] In the Indian context, the strengthening of castelessness developed alongside the success of a particular form of nationalism that opposed movements to destroy caste hierarchy and caste privilege.

The rise of castelessness also aligned with the strengthening of a Hindu nationalist ideology, which G. Aloysius describes as "communal nationalism in a double sense." Hindu nationalism, or Hindutva, rejected minority religions and oppressed castes—and for that reason Aloysius refers to it as "uppercaste Brahmanic nationalism."[69] It solidified as a formidable stream of nationalism in the 1920s through the writings of Vinayak Damodar Savarkar and the formation of the Rashtriya Swayamsevak Sangh (RSS, or the Association of the National Volunteers).[70] Savarkar, a Brahman from Maharashtra, charted the Hindu nationalist ideology with an anonymously published book in 1923 entitled *Hindutva: Who Is a Hindu?*[71] He emphasized the importance of claiming one's homeland within the physical territory of "Hindustan," of inheriting "Hindu blood," and of having one's holy land in Hindustan. Savarkar argued that because Muslims and Christians do not observe their holy land within Hindustan, they do not possess the essential characteristics of Hindutva. Christophe Jaffrelot summarizes Savarkar's motto as "Hindu, Hindi, Hindustan!" which involved "the superimposition of a religion, a culture, a language, and a sacred territory—the perfect recipe for ethnic nationalism."[72] Hindutva eventually became a cultural and political identity for the nation. While Savarkar critiqued Brahmans, he did not attack the ideological structure of caste (that is, Brahmanism), nor offer a critique of caste apartheid.

Hindutva spread as a grassroots movement through the RSS. Following a meeting between Savarkar and Keshav Baliram Hedgewar in the mid-1920s, Hedgewar founded the RSS in Nagpur in 1925 to propagate a Hindu nationalist ideology and physically strengthen the community through grassroots organizing. The RSS became the most powerful Hindutva organization with over 600,000 volunteers (*swayamsevàks*) by the time of independence from British rule.[73] Yet as the RSS spread in the 1930s and 1940s by creating local branches (*shakhas*) throughout the Bombay Presidency, Central Provinces, Punjab, and other parts of north India, the organization remained largely separate from formal politics. It initially spread among caste elites and became an organization of largely oppressor-caste Hindus. The RSS stayed out of formal politics during this period, but Savarkar and his ideology shaped Congress and the predecessor parties to the BJP.

Hindu nationalism had a strong presence within Congress during the 1920s and early 1930s and shaped competing schools of thought within the party. Members of the Hindu Mahasabha, who aligned with Savarkar's Hindutva, operated within Congress until 1937. As Gandhi's power rose within Congress, the lobbying efforts of this group weakened, as did the influence of Muslim League leaders.[74] Gandhi viewed the nation as inclusive of the diversity of communities that existed within the territory, while the members of the Hindu Mahasabha rejected Gandhi's version of Hinduism and his commitment to nonviolence. By 1937, Gandhi's appeal was global, and Hindu Mahasabha members became sidelined and decided to transform the Hindu Mahasabha into a separate political party under the leadership of Savarkar.[75] While Gandhi insisted that he and Congress represented all groups living within the borders of British India, many leaders of India's religious minorities and caste-oppressed groups did not experience Gandhi's philosophy as secular or inclusionary. At the same time, Hindu nationalists opposed Gandhi's efforts to reform Hinduism. The legitimacy of Congress's plan to be "all-India" representatives came into question as Muhammad Ali Jinnah and the All-India Muslim League rose in prominence as political representatives for Muslims and as Ambedkar and other Dalit leaders became political representatives of "untouchables."[76] Gail Omvedt describes how Hindu secularists, in the Nehruvian tradition, and Hindu reformists, in the Gandhian tradition, both identified Hinduism as the tradition of India.[77] In contrast, Ambedkar treated Hinduism and Buddhism as "contending schools of thought" from within India—the former authoritarian and the latter democratic.[78] Political philosopher Kancha Ilaiah Shepherd argues that Buddhism could have offered an "indigenous" political philosophy that challenged Brahmanism but that Ambedkar's writings on Buddhism remained underappreciated as a nationalist framework and failed to shape an anti-caste ideology within the mainstream nationalist movement.[79]

An explicitly Hindu nationalist political ideology became further institutionalized in predecessor parties that laid the foundation for the BJP and its co-adoption of Hindutva and castelessness. Savarkar asked the RSS to support the Hindu Mahasabha following its split from Congress in 1937. The RSS leadership, under M. S. Golwalkar, who was then a senior leader and the supreme leader (*sarsanghachalak*) from 1940 to 1973, initially refused.[80] Things began to shift a decade later, after a former RSS volunteer killed Mahatma Gandhi in 1948, less than six months after the bloody partition that split the colonial territory of "British India" into independent India and Pakistan. Hindu nationalists suffered a setback as prime minister Jawaharlal Nehru imposed a ban on the organization and had 20,000 *swayamsevak*s arrested.[81]

RSS leaders decided they needed to align with a political party that could advocate on their behalf. Golwalkar met with the president of the Hindu Mahasabha, and the leadership of both organizations created the Bharatiya Jana Sangh in 1951. RSS leaders helped get the party up and running, and the Jana Sangh (as it became widely known) was a forerunner to the BJP.[82] A more moderate stream within the Jana Sangh eventually led it to merge with other parties to form the Janata Party in 1977. After the Janata Party leadership banned its members from also belonging to the RSS, former Jana Sangh members left the Janata Party and created the BJP in 1980.[83] The BJP and RSS have maintained a close relationship over the past forty years, with grassroots mobilization by the RSS contributing to the BJP's rise to power as the dominant political party in the twenty-first century. Key to the BJP's ascendancy has been the mobilization of a caste-elite Hindu cultural and religious identity as the Indian national identity. At the same time, the BJP's Hindutva has continued to minimize attention to caste-based privilege and power, promote an ideology of castelessness to keep caste-oppressed groups within the Hindu fold, and require religious minorities to limit signs of their faith to the private sphere—often enforced through violence and terror.[84] Hindutva has pushed to restore the "traditional" social order by sidelining the histories and contributions of Muslims and caste-oppressed groups, advocating for policies and social norms that exclude "non-indigenous" religious groups, minimizing the anti-Brahmanical foundations of Buddhism and other indigenous religions that offer a powerful critique of Hinduism, and justifying discriminatory practices against Muslims, Christians, Dalits, and other caste-oppressed groups.[85]

Political theorist Gopal Guru explains how for Dalits democratic institutions have been the "initial rather than the sufficient condition" to realize "their normative goals."[86] Ambedkar mistrusted the political elite to consider the economic or educational needs or address the structural barriers and indignities faced by Dalits and other caste-oppressed groups, and tried to put economic safeguards in the Constitution and dismantle caste hierarchy by shaping Congress policies as India's first law minister.[87] Congress, instead, took the position that untouchability should end while caste could continue to structure marriage and family, religious customs, traditions, and social life in the familial, social, and religious realms. Yet enforcement against those who practiced untouchability was negligible, and there was little effort to dismantle caste hierarchy. While many laws—such as land reform efforts—had the veneer of equality, the state designed and implemented them in such a way that they brought limited improvements to caste-oppressed groups and continued to perpetuate a caste-based system of land access and ownership.[88] Satish Deshpande argues that Congress remained in power at the center for more than three decades seeking "to exorcise the embarrassment of caste" by "valorising a worldly ideal of castelessness."[89] The muting of caste identities by political leaders was not accompanied by similar efforts to promote anti-caste policies, ideologies, and practices to dismantle caste as a source of privilege. Brahmanism remained embedded in political institutions after the Congress leadership took a conservative position to maintain caste in social and economic life and not root out Brahmanism in key institutions of society. Instead, new mechanisms to hoard resources and opportunities have emerged.

Castelessness has become aligned with streams of nationalism propagated by Congress and the BJP. The universal "Indian" identity is inaccessible for many, including Dalits, Adivasis, so-called *lower* castes, and Muslims—whose caste and religious identities are highly visible and dictate their treatment and access to resources and opportunities.[90] Castelessness and Hindutva have become firmly institutionalized in the Indian state. Discussing legal scholar Marc Galanter's research,[91] Satish Deshpande explains:

> … the biggest boon that the state grants to the upper castes is a guarantee of anonymity in caste terms. This effectively means that regardless of the extent of their past or present privileges, their caste identity can never be used directly to prohibit or limit access to any public resource. In other words, the upper castes cannot be prevented from cornering a disproportionate share—or even all—of a public resource because they belong to caste A or B; their share can be limited only by setting aside portions exclusively marked for castes X and Y.[92]

Given their networks and resources, caste elites in independent India have been able to continue to hoard resources and dominate knowledge-making processes, yet these patterns are not visibly connected to caste privilege but are instead justified by "individual success and merit."[93] Only the resources that state policy explicitly reserved for historically oppressed groups remained directly inaccessible to caste elites. Castelessness has silenced structural advantages accumulated across generations and caste-based mechanisms of material and social advantage. As such, a caste-wise enumeration in the census would reveal the extent to which historical and ongoing caste privilege structures the accumulation of resources and power.

At the same time, the promise of democracy and the expansion of egalitarianism and civil society have flowed from the labor, politics, and organizing of caste-oppressed groups and other marginalized communities.[94] Sociologist Suryakant Waghmore explains how social transformation is "not achieved through the give of a benevolent state or the product of elite self-reform as Gandhi envisaged but sustained struggles of lesser citizens like those of Dalits."[95] Institutions of power have reconfigured often with new methods of retaliation in response to organizing by oppressed groups to dismantle gendered caste hierarchy. The decennial censuses of independent India have become a site of struggle over the invisibility of caste privilege, yet have fully aligned with dominant streams of nationalism that have remained acritical of caste and the otherizing of Muslims—allowing for the further institutionalization of castelessness and Hindutva within political institutions.

The Politics of Census Making

Social scientists and historians who study the production of data argue that censuses are neither neutral counts of populations nor matters of bureaucratic routine. Challenging the view that census data represent "social facts," scholars have argued that the entire process of conducting a census—from the development of the survey instrument to the compilation of collected data—is deeply political. Census data reflect a particular historical context and are embedded in contested visions of nationhood.[96] Sociologist Tianna Paschel states in her research on Colombia that "behind each estimate of the black population lies a complex and contentious political process."[97]

Scholars have historically situated the spread of national censuses as part of broader processes of rationalization, standardization, quantification, statistical thinking, surveillance, and disciplining and control, and theorized the role of censuses in the colonization of territories and peoples, the centralization of state power, the expansion of state capacity, and the creation of modern nation-states.[98] Censuses are part of the administrative practices through which modern states exercise symbolic power, which sociologist Pierre Bourdieu describes as the power to "constitute the given."[99]As an important tool in state administration, censuses classify people into groups to make local social relations legible to a centralizing state and jointly create subjects and knowledge.[100] Sociologist Bruce Curtis describes *census making* as "the process of identifying political subjects and centralizing knowledge."[101]

In the Indian subcontinent, caste was a central category of enumeration in colonial censuses. Research on colonial censuses suggests that the actual process of producing census data on caste was negotiated and political, and that the collection of caste data contributed to the hardening of caste boundaries.[102] Censuses were tools of colonialism and extractive capitalism during British rule, but also helped to recognize political subjects who actively demanded that the colonial state address gross inequities and violence related to caste hierarchy. The next chapter looks more closely at the enumeration of caste in colonial censuses. Here, I highlight five insights from the comparative research that inform this book's view of census making: (*a*) census making requires an extensive infrastructure that shapes the production of data; (*b*) commensuration is a social process; (*c*) material objects play an active role in census making, and these technologies are not neutral; (*d*) enumeration and classification are related yet analytically distinct; and (*e*) decisions to enumerate and classify caste-like systems are deeply embedded in historically specific understandings of citizenship and nation, and are shaped by bureaucratic practices, powerful ideologies of inclusion and exclusion, and social movements.

An extensive apparatus develops in the design and implementation of a census. Senior government administrators, census bureau officials, and other experts help envision the project from the design of the household instrument to the process of collecting, analyzing, and circulating data. The collection of data on social categories is more than just a technocratic act of enumeration; it is a political act entrenched in internal bureaucratic negotiations and affected by external organizing. Political scientist Melissa Nobles shows in her historical study of the United States (US) and Brazil how census agencies are political actors and "not the detached registers they purport to be."[103] State and local government agencies and private actors play a key role in implementing censuses. An extensive network of people translates the policies and intentions of centralized planners into practice and helps to secure participation from local communities and households. This "infrastructural work" is the taken-for-granted "laborious enterprise" of conducting a census that "requires substantial investment and ingenuity" by local actors.[104] Infrastructural work emerges from the interplay of centralized decisions, on-the-ground resources, and local histories, politics, and worldviews. Census making also involves households sharing information and consenting to data collection. Local actors do much of the *consent-building*, or what sociologist Michael Rodríguez-Muñiz describes as the on-the-ground labor of creating legitimacy for the census, mobilizing households to participate, and making marginalized groups legible.[105] Infrastructural work shapes processes of consent-building (including households' knowledge or perception of censuses and willingness to participate), the production of data during interviews, and the compilation and eventual dissemination of census data. Where censuses are still administered face-to-face, enumerators bring "to life" questions and answer options during interviews. Sociologists Mara Loveman and Jerônimo Muniz find in their study of censuses in Puerto Rico that enumerators played an active role in the construction of racial data during the first half of the twentieth century. As intermediaries between households and the US Census Bureau, and as local social elites, the enumerators' perceptions and worldviews shaped the classification of individuals by race. Their interviews and the resulting data led to a dramatic "whitening" of the colony across successive decennial censuses in response to the perceived and actual costs of being seen as nonwhite by the US colonial state.[106] Contestations within the centralized state bureaucracy, across levels of government, and between state and society actors shape census making and the creation of official state knowledge.

The production of census data involves *commensuration*. Sociologists Wendy Espeland and Mitchell Stevens define commensuration as "the transformation of different qualities into a common metric."[107] It involves the delineation of a boundary around something that we wish to value and the creation of a common metric or set of comparable categories to encompass everything within the boundary.[108] Commensuration is "crucial to how we categorize and make sense of the world" as it allows for "people to quickly grasp, represent, and compare differences."[109] While censuses may appear to simply "be a method for counting people," they are in fact "a mechanism for constructing and evaluating relations among citizens of a state or region"—a perfect example of commensuration.[110] A particular concept of citizenship, or of shared life within the boundaries of a particular territory, motivates the work of standardizing and quantifying "the relations among diverse peoples" in a census.[111] Commensuration describes decisions that we often deem only as "technical," but it is a deeply social and political process that involves considerable work and has powerful consequences. Commensuration is power, as it involves "creating social categories and backing them with the weight of powerful institutions."[112] In doing so, the social process of commensuration may also erase or minimize some identities and groups while creating or elevating others.[113]

Material objects and technologies also play an important role in the production of census data. The social and material worlds are deeply intertwined in the production of scientific knowledge, as leading scholars in science and technology studies have shown in ethnographic accounts of laboratories and surveys.[114] The census schedule, enumerator guide, data-entry tablets, and data-entry program are all examples of material objects that link the human actors involved in censuses and perform an under-examined role in the production of state data. In addition, computing technologies are not neutral; scholars of critical caste and technology studies, such as Palashi Vaghela, have shown how supposedly apolitical technologies and tech-related organizations embed and reproduce Brahmanical values.[115]

Fourth, processes of classification and enumeration are interrelated yet analytically distinct—and they are part of the broader work of commensuration. In his research on colonial censuses in Canada, Bruce Curtis defines classification as "the grouping of subjects together to form a 'population' whose elements may then be selectively disaggregated and made the objects of social policy and projects."[116] Classification involves the creation of categories that reflect historically specific conceptions of group relationships that emerge through contestations. Enumeration is closely related to classification yet focuses on the people, practices, and objects involved in the production of data. It involves the exchange between enumerators and respondents during the household interview (or related technologies that have replaced face-to-face enumeration), in which "bodies are counted, homogenized, and bounded in their extent" or in which classification takes place.[117]

While commensuration during censuses, and related processes of classification and enumeration are central to modern bureaucratic administration, groups have continuously contested if, how, and for what purpose national censuses should include questions on race, caste, gender, citizenship, and other sociopolitical categories. Whether a state's system of ethno-racial classification seeks to undergird "antidemocratic polities, segregated societies, and repressive, violent, even genocidal state projects" or "reconstruct historically exclusionary nations as inclusive political and cultural communities" remains an empirical question. Mara Loveman finds in her study of nineteen Latin American countries that "the politics and practices of demarcating categorical divides and naturalizing them as group boundaries are endogenous to the processes that generate racial inequality and injustice, not exogenous to them."[118] She shows how census officials are involved in national politics while engaging in international scientific debates about how to measure national progress and advance particular views of race. Census officials in Latin America played a key role in marshaling racial statistics to create and actualize a white-supremacist nation-building narrative of "progress" toward a whiter future.[119] Censuses in the Americas have been deeply embedded in the production of racial hierarchy through processes of state administration and the consolidation of political and economic power.[120] Loveman also argues that "the absence of official racial classification can itself be a powerful means of exclusion."[121] For example, in France, national censuses have excluded the enumeration of ethnicity since the nineteenth century. The decision to not enumerate ethnicity has been under scrutiny since the mid-1980s; immigrant organizations have highlighted the structural discrimination experienced by historically colonized immigrants and racialized groups and critiqued the political right's attack on immigrants in defense of a "pure" French nation. They have argued that census data on ethnicity could provide powerful information to document the extent of exclusion.[122]

When I interviewed a caste census activist in early 2013, he said, "You know you are talking to the losing side, right?" The SEC survey data collection was still ongoing in much of the country, but the activist knew that the extensive energy that he and his colleagues had expended in the lead-up to Census 2011 had not translated into their supposed win from May 2010. With a discouraged tone he explained, "Whoever has benefited most under the current system—they don't want to see anymore change."[123] This book's story of census making explores Loveman's assertion that the absence of data collection—in this case on caste—can itself be a powerful tool of exclusion. The decennial census of India obscures the concentration of caste-based resources and power by making invisible caste privilege. In a 1955 essay, Ambedkar apologizes for not being able to provide census data to support his description of the caste system and highlights the home minister's role in blocking the collection of caste-wise data in the first decennial census of independent India.[124] Now, almost seventy years later, this book traces how bureaucratic deflection operates to protect an ideology of castelessness within the census in response to demands for a policy change.

The next chapter, entitled "The Institutional Life of Caste," examines the broader relationship between Brahmanism and the state by focusing on caste in the decennial census. The first half of the chapter explores caste and censuses in the colonial period. Censuses were part of a Brahmanical sociology of knowledge that heavily shaped administrative understandings of caste.[125] Many British officials believed that data on religion and caste were "the sociological keys to understanding the Indian people" and these data fed into army recruitment, education policy, balancing the proportion of Hindus and Muslims in public service, and theories about how certain castes "were organizing to supplant British rule."[126] Yet the census, and more specifically the caste and religion questions, also became a site of resistance to Brahmanical power. I discuss Brahmanism's entrenchment in state institutions alongside anti-caste activists fighting against the institutionalization of caste power. The second half of the chapter focuses on caste in the censuses of independent India and argues that the decision to eliminate a caste-wise enumeration was consistent with the strengthening of ideologies of castelessness and Hindutva. Chapter 2 concludes by describing the unsuccessful attempt to include a caste-wise enumeration in Census 2001 even after the extension of central government affirmative action benefits to OBCs in the 1990s. The chapter draws upon government documents, judicial decisions, parliamentary debates, newspaper archives, public proceedings, and secondary sources.

Chapter 3, "The Politics of the Count," takes a macro view of the politics surrounding the collection of caste data in Census 2011. Activists organized a centralized campaign to include a caste-wise enumeration in the decennial census. They argued that caste data would help to quantify the degree to which caste-based inequalities persist and support the implementation of policies to create more egalitarian institutions. In response, political leaders made a historic decision to change census policy on caste. The chapter details the backlash that followed the public concession to include a caste-wise enumeration in the census and shows how bureaucratic deflection unfolded. Executive bureaucrats worked behind-the-scenes to prevent a change in census policy on caste. They eventually pushed the caste count into a state project with a history of producing poor-quality data. Multiple maneuvers reframed the debate as technical, minimized ideological differences, homogenized dissimilar types of caste consciousnesses, promoted an uncritical nationalism, elevated the importance of expertise, presented expertise as apolitical, shifted key decisions to bureaucrats, and compartmentalized anti-caste efforts to peripheral locations within the bureaucracy. Executive bureaucrats created leeway to redefine priorities and reframe the narrative, which they presented as protecting the "integrity of the census" from the "politics of caste." They backtracked on their public concession in the less visible spaces of the bureaucracy, where expertise dominates decision-making and civil society access is limited. As such, the role of Brahmanism in structuring expertise in the bureaucracy is difficult to detect. In tracing the back and forth between caste census activists and the executive bureaucracy, the chapter shows how bureaucratic deflection unfolds and how new mechanisms of Brahmanism emerge relationally. The chapter draws upon in-depth interviews, newspaper archives, and parliamentary debates to show how the central bureaucracy and allied institutions maintained the status quo in Census 2011.

Chapter 4, "Survey Making in an Era of Castelessness," follows the process of bureaucratic deflection as the MoRD begins to implement the caste-wise enumeration in an alternative state project. This chapter traces the centralized planning and management of the BPL survey that structured ground-level data collection operations throughout the country. I trace the widening gap between the demands made by caste census advocates and the implementation of caste-related questions within the revamped BPL survey. Several decisions created a process that differed from the one envisioned by caste census advocates: the contracting out of data-entry operations and data processing to the private sector, the use of tablet PCs for the first time in a nationwide survey, the creation of a two-person data collection team, the wording of answer options for the caste-related questions, and the decentralization of publicity for the nationwide survey. Project managers in Delhi decided the configuration of the data collection team and what questions to ask and answer options to include. While the government and national media heralded the use of tablets during the household interview as a sign of "progress and development," it diverted attention away from the necessary planning and public education campaign for a successful nationwide enumeration of caste. Chapter 4 draws upon interview data with government officials and private contractors, observational data of enumerator trainings and household interviews, and a range of primary documents to trace how processes of bureaucratic deflection continued to unfold during the planning stage of the revised BPL survey-cum-caste-wise enumeration.

Chapter 5, entitled "The Household Interview," takes an ethnographic view of ground-level survey operations in the south Indian state of Karnataka. It follows the routine work of enumerators, data-entry operators, local bureaucrats, master trainers, and private contractors over the course of a year and a half as they collect caste-wise data. The chapter opens with a vignette of a household interview in Bengaluru. It highlights differences between the "imagined" interview—by project managers in Delhi—and the actual on-the-ground social process of collecting data from households. Drawing upon observations of over 350 household interviews (that is, 250 regular interviews and 100 "re-enumeration" interviews) and 29 post-survey interviews, as well as interviews of data collectors, this chapter focuses on practices that produce caste-wise data during household interviews. In doing so, it traces how earlier decisions by executive bureaucrats in Delhi structured the day-to-day work of data collectors. I also find that the data-entry operators' understandings of caste and religion disproportionately shaped the production of caste-wise data.

This chapter traces how centralized policies and planning intersect with local histories to shape the production of caste-wise data. Appendix B describes the data collection and project methodology in greater detail.

Chapter 6, "Disappeared Data: The Life and Death of Caste Data," finishes the narrative by describing what happened to the collected caste-wise data. Drawing upon interview, observational, and document data, the chapter opens by following the processing and finalization of collected data in Karnataka and the national story of (not) analyzing, compiling, and publishing the caste-wise data. Finalizing the data involved recollecting data in select neighborhoods, reviewing the collected data, and separating the caste-wise data from the rest of the BPL data (including splitting the SC/ST data from the other caste-wise data). While the MoRD tabulated, analyzed, and eventually published the bulk of the data from the SEC survey in 2015, and the MoHUPA in 2016, the caste-wise data (except for the SC/ST data) remained untouched on a government server. The BJP-led government, in power since 2014, faced political pressure to release the caste-wise data after the MoRD published the data for BPL identification. In response, the political leadership announced the creation of an expert committee in July 2015 to analyze the caste-wise data. However, the committee never published the caste-wise data, and it remains unclear if the committee ever met. The political leadership buried the data.

Chapter 7, the conclusion, entitled "Commensuration in Brahmanical Institutions," reviews the theoretical contributions of the manuscript and contextualizes the key findings of the book. I revisit the politics of bureaucratic deflection and discuss a recent challenge to castelessness in Bihar. This chapter discusses the importance of, and likely challenges to, future efforts to collect and publish caste-wise data in the decennial census.

While this book focuses on the contemporary Indian case, it also draws upon a rich historical and comparative body of research on states, censuses, caste-like systems, and the expansion of racial and colonial capitalism. For more than 150 years, anti-caste scholar-activists have traced parallels between caste–gendered hierarchy in India and systems of racialized and gendered oppression across the globe.[127] This book is inspired by a tradition of writers whose historically grounded and place-specific research on caste emerged alongside an interest to understand caste-like systems in other places and stand in solidarity with those working to dismantle systems of domination in their own communities throughout the world.[128] Core to the mission of sociological research—as laid out by W. E. B. Du Bois in the United States and

B. R. Ambedkar in India—is to make visible oppressive structures, ideologies, and subjectivities in order to create egalitarian institutions and support struggles for liberation. During a period when caste-based violence and Hindutva have flourished in prime minister Modi's Hindu nationalist state, this book seeks to highlight how bureaucratic deflection prevents the implementation of policies that strive to make caste-based power visible behind seemingly "modern" ideologies of castelessness, expertise, and nationalism.

Notes

1. Interview with activist, June 17, 2016.
2. Sunil Prahbu, "Pranab Says Current Census Will Include Caste, after All," NDTV, May 7, 2010; HT Correspondent, "Pranab Justifies Caste-Based Census," *Hindustan Times*, May 8, 2010; "Lok Sabha (No. 15) Debate" (New Delhi: Lok Sabha Secretariat, May 7, 2020), 570.
3. HT Correspondent, "Pranab Justifies Caste-Based Census."
4. Government caste and tribe categories divide the population into four major all-India groupings: Scheduled Castes (SCs), Scheduled Tribes (STs), Other Backward Classes (OBCs), and the "general" category. In the current day, SCs consist of Dalits (ex-"untouchables"), who identify as Hindu, Buddhist, or Sikh; Dalit Muslims and Dalit Christians are not considered SCs. Indigenous communities, or Adivasis, belonging to any religious background form the ST category. OBCs include diverse groups, from Shudras (or peasant subcastes) to Dalit Muslims and Dalit Christians, and other groups that vary by region. The "general" category consists of caste elites or caste-privileged groups, such as Hindu Brahmans, Ashraf Muslims, and Syrian Christians, although there are differences across states. Colloquially, the term "*lower* caste" is often used to refer to Shudras, or in other instances Dalits, Adivasis, and OBCs, while Bahujan also refers to these three groups and religious minorities (particularly Muslims). Caste as *jati* refers to the idea of lineage or to a kinship group that is identified in a local setting; families marry within their *jati*. There are thousands of *jati*s in India. Each state administrative category consists of hundreds of *jati*s.
5. B. R. Ambedkar also resigned because of the executive bureaucracy's resistance to implement policies and promote legislation to destroy caste and patriarchy and expand protections for SCs and STs. Ambedkar was a leading activist, prolific scholar, and political leader in the twentieth century who sought to destroy caste and patriarchy and empower Dalits and other oppressed groups. He was chair of the drafting committee of the Indian Constitution and the first law minister of independent India, despite not being a member of the Congress Party. Ambedkar opposed the government's decision to remove a full caste-wise enumeration from the first census of independent India in 1951.

B. R. Ambedkar, "Statement by Dr. B. R. Ambedkar in Parliament in Explanation of His Resignation from the Cabinet (Annexure 1)," in *Dr. Babasaheb Ambedkar: Writings and Speeches, Vol. 14, Part 2*, ed. Vasant Moon, 1315–27 (New Delhi: Dr. Ambedkar Foundation, Ministry of Social Justice & Empowerment, Government of India, 2014); B. R. Ambedkar, "Thoughts on Linguistic States [1955]," in *Dr. Babasaheb Ambedkar: Writings and Speeches, Vol. 1*, ed. Vasant Moon, 138–201 (New Delhi: Dr. Ambedkar Foundation, Ministry of Social Justice & Empowerment, Government of India, 2014), 167.

6. More specifically, in the *Indra Sawhney & Others v. Union of India* case, the Supreme Court of India decided that 27 percent of central government reservations for OBCs were valid and introduced the "creamy layer" criterion, such that OBCs above a specified income criteria are ineligible for reservations.
7. The estimated size of the OBC population varies. The Mandal Commission estimated an OBC population of 52 percent in 1980, while the National Sample Survey (NSS) estimated an OBC population of 36 percent in 1999–2000, 41 percent in 2004–2005, and 44 percent in 2011–2012. See Backward Classes Commission (B. P. Mandal, Chair), *Report of the Backward Classes Commission* (Delhi: Government of India, 1980); Chetan Chauhan, "Updated OBC Count: 41 Pc," *Hindustan Times*, November 1, 2006; Sonalde Desai, "Reimagining the OBC Quota," *The Hindu*, September 19, 2017, sec. Opinion.
8. I cite contemporary scholars Kancha Illaiah Shepherd and Surinder Jodhka, who build upon Ambedkar's view of caste as a system of "graded inequality" arranged according to an "ascending scale of reverence and a descending scale of contempt." Kancha Ilaiah Shepherd defines caste as a "Brahmanical institution based on *varnadharma*. The modalities of all other institutions like marriage and family are structured by Brahmanical ideologies to perpetuate caste." B. R. Ambedkar, "Who Are the Shudras? How They Came to Be the Fourth Varna in Indo-Aryan Society (1947)," in *Dr. Babasaheb Ambedkar: Writings and Speeches, Vol. 7*, ed. Vasant Moon, 3rd ed., 5–227 (New Delhi: Dr. Ambedkar Foundation, Ministry of Social Justice & Empowerment, Government of India, 2020); Kancha Illaiah Shepherd, *God as Political Philosopher: Buddha's Challenge to Brahmanism*, 2nd ed. (New Delhi: Sage, 2019), 23; Surinder Jodhka, *Caste in Contemporary India* (New Delhi: Routledge, 2016).
9. Household amenities refer to the availability and source of drinking water, source of lighting, availability and type of latrine, type of wastewater connection, availability of bathing facilities and kitchen, and type of fuel for cooking. Household assets include whether or not the household has a radio, television, computer, mobile, bicycle, scooter, car, and available banking services. Household data were collected during the household listing phase (or first round) of Census 2011, which started in May 2010.

10. HT Correspondent, "Write Meri Jaati Hindustani in 2011 Caste Based Census," *Hindustan Times*, May 16, 2010; HT Correspondent, "Civil Society Movement Opposes Caste Census," *Hindustan Times*, June 11, 2010; Staff Reporter, "Movement against Caste-Based Census Launched," *The Hindu*, June 11, 2010; "Movement against Inclusion of Caste in Census Gains Impetus," *The Hindu*, June 12, 2010. For analysis of the media coverage, see Dilip Mandal, "Upper-Caste Domination in India's Mainstream Media and Its Extension in Digital Media," *Economic and Political Weekly* 55, no. 46 (November 21, 2020): 34–39.
11. Sujay Mehdudia and Siddharth Varadarajan, "Government Not for Caste Census," *The Hindu*, May 5, 2010; J. Venkatesan, "Moily Defends Caste-Based Census," *The Hindu*, May 21, 2010.
12. The government's use of the word "census" in this instance was imprecise. While the executive leadership renamed the BPL survey the Socio-Economic Caste (SEC) Census, I refer to the project as the SEC survey throughout this book because the revamped BPL survey did not fall under the purview of the Census Act.
13. The BPL survey had previously been conducted in 1992, 1997, and 2002. Researchers critiqued the 2002 BPL survey for data quality and coverage, corruption, and poor survey design. Scholars critiqued that many of the poorest rural households did not have a BPL card and that exclusion of poor households from BPL status increased among SCs, STs, day laborers, landless households, and those that lacked a connection to a local political leader. See K. Sundaram, "On Identification of Households Below Poverty Line in BPL Census 2002: Some Comments on the Proposed Methodology," *Economic and Political Weekly* 38, no. 9 (March 1, 2003): 896–901; Indira Hirway, "Identification of BPL Households for Poverty Alleviation Programmes," *Economic and Political Weekly* 38, no. 45 (2003): 4803–8; Sachin Kumar Jain, "Identification of the Poor: Flaws in Government Surveys," *Economic and Political Weekly* 39, no. 46 (November 20, 2004): 4981–84; Neela Mukherjee, "Political Corruption in India's Below the Poverty Line (BPL) Exercise: Grassroots' Perspectives on BPL in Perpetuating Poverty and Social Exclusion" (Development Tracks in Research, Training & Consultancy, New Delhi, 2005); Jyotsna Jalan and Rinku Murgai, "An Effective "Targeting Shortcut"? An Assessment of the 2002 Below-Poverty Line Census Method," World Bank, mimeo, 2007; Sabina Alkire and Suman Seth, "Determining BPL Status Some Methodological Improvements," *Indian Journal of Human Development* 2, no. 2 (2008): 407–24; Madhura Swaminathan, "Public Distribution System and Social Exclusion," *The Hindu*, May 7, 2008; Santosh Mehrotra and Harsh Mander, "How to Identify the Poor? A Proposal," *Economic and Political Weekly* 44, no. 19 (May 9, 2009): 37–44; Government of India, *Report of the Expert Group to Advise the Ministry of Rural Development on the Methodology for Conducting*

the Below Poverty Line (BPL) Census for the 11th Five-Year Plan (N.C. Saxena, Chair) (New Delhi: Ministry of Rural Development, Government of India, August 2009); Jean Drèze and Reetika Khera, "The BPL Census and a Possible Alternative," *Economic and Political Weekly* 45, no. 9 (February 27, 2010): 54–63; Indrajit Roy, ""New" Lists for "Old": (Re-) Constructing the Poor in the BPL Census," *Economic and Political Weekly* 46, no. 22 (May 28, 2011): 82–91; Sabina Alkire and Suman Seth, "Identifying BPL Households: A Comparison of Methods," *Economic and Political Weekly* 48, no. 2 (January 12, 2013): 49–57; Sabina Alkire and Suman Seth, "Selecting a Targeting Method to Identify BPL Households in India," *Social Indicators Research* 112 (2013): 417–46; Sitakanta Panda, "Political Connections and Elite Capture in a Poverty Alleviation Programme in India," *The Journal of Development Studies* 51, no. 1 (2015): 50–65.

14. Only the SC/ST data from the revised BPL (that is, the SEC survey) have been published. Previous BPL surveys also collected household level SC/ST data in rural communities.
15. Scholarship in this area roughly argues that while the significance of caste is declining in some areas of life, the role of caste in politics remains strong. Rajni Kothari, ed., *Caste in Indian Politics* (Delhi: Orient BlackSwan, 1995); Andre Beteille, "The Reproduction of Inequality: Occupation, Caste and Family," *Contributions to Indian Sociology* 25, no. 1 (1991): 3–28; Mysore Narasimhachar Srinivas, "An Obituary on Caste as a System," *Economic and Political Weekly* 38, no. 5 (February 1, 2003): 455–59; Dipankar Gupta, "Caste and Politics: Identity over System," *Annual Review of Anthropology* 34 (2005): 409–27.
16. The entry of politicians from historically oppressed groups into formal politics has challenged the configuration of political power at the regional and national levels. Anti-caste movements and the mobilization of oppressed castes have led to the diversification of elected leaders and improved access to public resources in many regions, although structural barriers and discriminatory systems continue. A. Kalaiyarasan and M. Vijayabaskar, *The Dravidian Model: Interpreting the Political Economy of Tamil Nadu* (Cambridge: Cambridge University Press, 2021); Nirmal Singh, *Dalits in Indian Democracy* (Delhi: Kalpaz Publications, 2018); Jeffrey Witsoe, *Democracy against Development: Lower-Caste Politics and Political Modernity in Postcolonial India* (Chicago: University of Chicago Press, 2013); Eleanor Zelliot, *Ambedkar's World: The Making of Babasaheb and the Dalit Movement* (New Delhi: Navayana, 2013); Christophe Jaffrelot and Sanjay Kumar, *Rise of the Plebeians? The Changing Face of the Indian Legislative Assemblies* (Delhi: Routledge, 2012); Christophe Jaffrelot, *India's Silent Revolution: The Rise of the Lower Castes in North India* (New York: Columbia University Press, 2003); Yogendra Yadav, "Reconfiguration in Indian Politics: State Assembly Elections, 1993–95," *Economic and Political Weekly* 31, no. 2/3

(January 13, 1996): 95–104; Yogendra Yadav, "Understanding the Second Democratic Upsurge," in *Transforming India: Social and Political Dynamics of Democracy*, ed. Francine Frankel, Zoya Hasan, Rajeev Bhargava, and Balveer Arora (Delhi: Oxford University Press, 2002), 89–119.

17. The recent scholarship on caste and politics minimizes attention to the role that executive bureaucrats, government officials, and other experts play in actualizing the demands made by citizens via their elected leaders. Some notable exceptions include: Suraj Yengde, *Caste Matters* (Delhi: Penguin Random House India Private Limited, 2019); Witsoe, *Democracy against Development*; Akhil Gupta, *Red Tape: Bureaucracy, Structural Violence, and Poverty in India* (Durham: Duke University Press, 2012); Gopal Guru, "Liberal Democracy in India and the Dalit Critique," *Social Research: An International Quarterly* 78, no. 1 (2011): 99–122; Sebastiaan Maria van Gool, "Untouchable Bureaucracy: Unrepresentative Bureaucracy in a North Indian State" (doctoral thesis, Leiden, The Netherlands, Leiden University, 2008).
18. Gianpaolo Baiocchi and Ernesto Ganuza, *Popular Democracy: The Paradox of Participation* (Stanford University Press, 2020); Caroline W. Lee, Michael McQuarrie, and Edward T. Walker, eds., *Democratizing Inequalities: Dilemmas of the New Public Participation* (New York University Press, 2015); Mark B Brown, *Science in Democracy: Expertise, Institutions, and Representation* (Cambridge, MA: MIT Press, 2009); Harry Collins and Robert Evans, *Rethinking Expertise* (Chicago: University of Chicago Press, 2007); Timothy Mitchell, *Rule of Experts: Egypt, Techno-Politics, Modernity* (Berkeley: University of California Press, 2002); Sheila Jasanoff, "(No?) Accounting for Expertise," *Science and Public Policy* 30, no. 3 (2003): 157–62; Miguel Angel Centeno, *Democracy within Reason: Technocratic Revolution in Mexico* (University Park, PA: Penn State Press, 2010).
19. Bernard Magubane, *The Making of a Racist State: British Imperialism and the Union of South Africa, 1875–1910* (Trenton, NJ: Africa World Press, 1996), 108, quoted in Thenmozhi Soundararajan, *The Trauma of Caste: A Dalit Feminist Meditation on Survivorship, Healing, and Abolition* (Huichin, unceded Ohlone land, aka Berkeley, CA: North Atlantic Books, 2022), 37.
20. A special thanks to Zophia Edwards and Orly Clergé for their feedback that highlighted the need to define and expand this concept.
21. Edwin Amenta et al., "The Political Consequences of Social Movements," *Annual Review of Sociology* 36 (2010): 287–307; Marco G. Giugni, "Was It Worth the Effort? The Outcomes and Consequences of Social Movements," *Annual Review of Sociology* 24 (1988): 371–93. For examples of research that focuses on the implementation stage, see Christopher L. Gibson, "The Consequences of Movement Office-Holding for Health Policy Implementation and Social Development in Urban Brazil," *Social Forces* 96, no. 2 (2017): 751–78; Kenneth T. Andrews, "Social Movements and Policy

Implementation: The Mississippi Civil Rights Movement and the War on Poverty, 1965 to 1971," *American Sociological Review* 66 (2001): 71–95.

22. Jaffrelot, *India's Silent Revolution*.
23. Backward Classes Commission (K. Kalelkar, Chair), *Report of the Backward Classes Commission* (Simla: Government of India, 1955), 170.
24. B. R. Ambedkar, "States and Minorities," in *Dr. Babasaheb Ambedkar: Writings and Speeches, Vol. 1*, ed. Vasant Moon, 381–449 (New Delhi: Dr. Ambedkar Foundation, Ministry of Social Justice & Empowerment, Government of India, 2014).
25. Galanter, *Competing Equalities*, 86.
26. Following the Supreme Court decisions *State of Madras v. Champakam Dorairajan* (1951) and *Venkataramana v. State of Madras (1951)*, parliament passed the First Amendment Act of 1951 to revise Articles 15 and 16 of the Constitution in order to specify that the "right to equality" did not prevent "special provision for the advancement of any socially and educationally backward classes of citizens or for the Scheduled Castes and the Scheduled Tribes." It reversed the Supreme Court decisions and enabled caste-based reservations. The initial legislation allowed reservations for a period of ten years. Since that time, every ten years, parliament has passed legislation in support of the extension of reservation. In these instances, other branches of government served as a check against the judiciary. The courts have played a major role in shaping policy by defining the constitutional boundaries of preferential treatment and affirmative action. Galanter, *Competing Equalities*, 382.
27. Galanter, *Competing Equalities*, 359.
28. Ibid.
29. I use the Group A classification to refer to both Group A and Class I. The original "Class I–IV" classifications of the administrative tiers in were replaced by "Groups A–D" classifications in the 1990s: Class I/Group A (senior administrative), Class II/ Group B (other administrative), Class III/Group C (clerical), Class IV (attendants, peons). Government of India Ministry of Personnel, Public Grievances and Pensions, "Brochure on Reservation for SCs, STs & OBCs in Services," 2014, 9–10. Central government jobs—even in the lower tiers of the bureaucracy—provided a stable income, government housing, and an important pathway for children and family members to access education and occupational mobility previously inaccessible to caste-oppressed groups.
30. While reservations in promotions started in 1955, they were largely limited to the lower tiers of the bureaucracy until 1974. The government also granted SC and ST civil servants two extra years to pass an exam required for promotion in 1975. Kathryn Victoria Bahnken Doner, "Seventy Years Later: Caste in the Indian Bureaucracy," *Sociology Between the Gaps: Forgotten and Neglected Topics* 7 (2022).

31. As mentioned at the start of the introduction, the government only implemented OBC reservations in the 1990s. By 2013, OBCs had 17.3 percent representation in central services—approximately 10 percentage less than their reservation quota. OBCs were also concentrated in group C jobs, with only 8.4 percent OBC representation within group A jobs. Government of India Ministry of Personnel, Public Grievances and Pensions, *Annual Report 2013–2014* (2014).
32. Government of India, Ministry of Personnel, Public Grievances and Pensions, *Annual Report 2016–2017* (2017).
33. "Rajya Sabha Unstarred Question No. 2686 by Dr. John Brittas, Answered by Ministry of Personnel, Public Grievances and Pensions," March 23, 2023.
34. The delayed implementation of reservations for students in India's most elite Indian Institute of Technology (IIT) and Indian Institute of Management (IIM) institutions, the comparatively late implementation of reservations for teaching/faculty positions in colleges and universities, and the related inhospitable and violent climate that caste-oppressed students experience in many colleges and universities speak to the slow implementation of efforts to diversify higher educational institutions.
35. Ashok Danavath, "A Discriminatory Education Policy That Further Excludes the Oppressed from Academia: The Case of the National Overseas Scholarship (NOS) for SC-ST Scholars in India" (Rapoport Center Human Rights Working Paper Series, The University of Texas at Austin, August 2023); Ashok Danavath and Jyoti Bania, "Why Aren't Enough Students from Tribal Communities Receiving the National Overseas Scholarship?" *The Wire*, July 27, 2021. Ashok Danavath also studied an overseas scholarship for caste-oppressed students administered by the state of Telangana since 2016 and found that funds have not been allocated to 300 students a year as originally announced, the budget has been cut in half from its initial amount, and money from this scholarship has been diverted to Economically Weaker Sections (EWS). Ashok Danavath, "Telangana's Backward Caste Overseas Scholarship Hampered by Inclusion of EWS Reservation Policy," *News Click*, April 27, 2022.
36. Sukhadeo Thorat, *Dalits in India: Search for a Common Destiny* (Delhi: Sage, 2009).
37. Danavath, "A Discriminatory Education Policy That Further Excludes the Oppressed from Academia."
38. Ibid.
39. Ibid.
40. Ibid. Ambedkar completed graduate studies at Columbia University in the United States, with sponsorship from Maharaja Sayaji Rao of Baroda.
41. G. Aloysius, *Nationalism without a Nation in India* (Delhi: Oxford University Press, 1997), 2.
42. G. P. Deshpande, "Introduction," in *Selected Writings of Jotirao Phule*, ed. G. P. Deshpande, 1–21 (New Delhi: LeftWord Books, 2002), 6–7.

43. G. P. Deshpande, "Introduction: On Infant Marriage and Enforced Widowhood," in *Selected Writings of Jotirao Phule*, ed. G. P. Deshpande, 191–92 (New Delhi: LeftWord Books, 2002), 191.
44. Jotirao Phule, "Slavery (1873)," in *Selected Writings of Jotirao Phule*, ed. G. P. Deshpande (New Delhi: Leftworld Books, 2002), 29.
45. B. R. Ambedkar, "Annihilation of Caste," in *Dr. Babasaheb Ambedkar: Writings and Speeches, Vol. 1*, ed. Vasant Moon 23–98 (New Delhi: Dr. Ambedkar Foundation, Ministry of Social Justice & Empowerment, Government of India, 2014), 72.
46. Phule and Ambedkar clearly distinguish between Brahmans and Brahmanism in their writings. G. Deshpande explains, "[W]hile Phule is unforgiving in his attack on Brahmanism, he is not against brahmans per se." G. P. Deshpande, "Introduction: On Infant Marriage and Enforced Widowhood," 191. Gail Omvedt describes Ambedkar's similar position when she cites his differences with non-Brahman leaders whom he describes as being "against Brahmans but not against Brahmanism, whereas I am against Brahmanism but not necessarily against Brahmans." Gail Omvedt, *Cultural Revolt in a Colonial Society* (Mumbai: Scientific Socialist Education Trust, 1976; New Delhi: Manohar, 2023), 9-10. Citation refers to the Manohar edition.
47. A second more narrow conceptualization of Brahmanism is also important to document. It emerges from non-Brahman movements in the Madras and Bombay presidencies and in princely Mysore (that is, in the southern and western peninsular regions) in the late nineteenth and early twentieth centuries. Each regional movement had a distinct history, yet they did not emerge in isolation and consistently organized against the power of Brahmans, who controlled access to key institutions of society. While these movements built off the earlier writings and activism of Jotirao Phule and Iyothee Thass, among others, some within them tended to have a narrower view of Brahmanism—focused on the privileges and power of Brahmans—in contrast to the ideology that undergirded caste hierarchy. Non-Brahman movements sought to open access to schooling, government jobs, public spaces, and local elected bodies. As these movements grew, locally dominant non-Brahman groups disproportionately shaped the demands and approaches of the movements; they had considerable advantage in accessing reserved seats in government jobs and scholarships for schooling due to their higher rates of education, English literacy, and inherited economic, political, and social resources and networks. Non-Brahman movements were less effective in securing universal access to schooling, land, small business ownership, and a wider range of employment opportunities. Still, where anti-caste organizing occurred, including non-Brahman and anti-Brahman movements, comparatively lower levels of inequality and higher levels of human and social development persist into the current day.

For more on non-Brahman movements, see Eugene Irschick, *Politics and Social Conflict in South India* (Berkeley and Los Angeles: University of California Press, 1969); Omvedt, *Cultural Revolt in a Colonial Society*; V. Geetha and S. V. Rajadurai, *Towards a Non-Brahman Millennium: From Iyothee Thass to Periyar* (Calcutta: Samya, 1998).

48. V. Geetha and S. V. Rajadurai, "Dalits and Non-Brahman Consciousness in Colonial Tamil Nadu," *Economic and Political Weekly* 28, no. 39 (September 25, 1993): 2091–98.
49. Phule argued against extreme exclusion and violence that Brahman widows faced within the Brahmanical structures of caste and patriarchy. Jotirao Phule, "Opinion from Jotteerao Govindrao Phulay on Note No. II, by Mr. B. M. Malabari on Enforced Widowhood," in *Selected Writings of Jotirao Phule*, ed. G. P. Deshpande, 195–197 (New Delhi: Leftworld Books, 2002), 195.
50. Sharmila Rege, *Against the Madness of Manu: B.R. Ambedkar's Writings on Brahmanical Patriarchy* (Delhi: Navayana, 2013), 20.
51. Ibid., 20, 62–63, 83.
52. Pavalabai was simultaneously condemned as vulgar and desired as a Tamasha artist, and unworthy of documentation during her lifetime unlike her performance partner. Pavalabai's partner, Bapurao, who was caste and male privileged, was constructed as decent and worthy of extensive documentation as a Tamasha artist. Shailaja Paik, *The Vulgarity of Caste: Dalits, Sexuality, and Humanity in Modern India* (Stanford, CA: Stanford University Press, 2022), 73–74.
53. In theorizing how Brahmanism operates, anti-caste and Dalit feminist scholars have highlighted how caste hierarchy and gender hierarchy are the "organizing principles of the Brahmanical social order." Historian Uma Chakravarti highlights that "as Ambedkar had pointed to caste as a system of graded inequalities, we should note that patriarchies in the subcontinent were contained within a larger system which was graded according to caste." Uma Chakravarti, "Conceptualizing Brahmanical Patriarchy in Early India: Gender, Caste, Class, and State," *Economic and Political Weekly* 28, no. 14 (1993): 579–85, 579. Historian Shailaja Paik critiques mainstream feminist discourses as being trapped "forms of Brahmanism," and political theorist Gopal Guru critiques how caste elites dominate Indian feminist theory and politics. Shailaja Paik, "Amchya Jalmachi Chittarkatha (The Bioscope of Our Lives): Who Is My Ally?," *Economic and Political Weekly* 44, no. 40 (October 3, 2009): 39–47, 45; Gopal Guru, "Dalit Women Talk Differently," *Economic and Political Weekly* 30, no. 41–42 (1995): 2548–50. For additional analyses of how patriarchy and caste are foundationally intertwined or how gender and caste co-constitute material conditions and power, see Vasanth Kannabiran and Kalpana Kannabiran, "Caste and Gender: Understanding Dynamics of Power and Violence," *Economic and Political Weekly* 26, no. 37

(1991): 2130–34; Sharmila Rege, *Writing Caste/Writing Gender: Reading Dalit Women's Testimonies* (Delhi: Zubaan, 2006); Uma Chakravarti, ed., *Thinking Gender Doing Gender: Feminist Scholarship and Practice Today* (Hyderabad: Orient BlackSwan, 2016); Uma Chakravarti, *Gendering Caste through a Feminist Lens* (Delhi: Sage, 2018); Sunaina Arya and Aakash Singh Rathore, eds., *Dalit Feminist Theory: A Reader* (London: Routledge, 2019); Sunaina Arya, "Dalit or Brahmanical Patriarchy? Rethinking Indian Feminism," *CASTE: A Global Journal on Social Exclusion* 1, no. 1 (2020): 217–28; Paik, *The Vulgarity of Caste*; Shailaja Paik, *Dalit Women's Education in Modern India: Double Discrimination* (London and New York: Routledge, 2014).

54. At the same time, the lower tiers of the administrative bureaucracy and military became important sources of government jobs for historically oppressed groups. As Shailaja Paik writes,

    > British rule unleashed modern forces that opened up new employment opportunities for Dalits. The British enabled Untouchables' access to law, education, and the political realm, yet such access was frustrated by ambiguities, uncertainties, and contradictions. For instance, the British also upheld religious restrictions, favored the dominant castes, and reinforced caste ideology.

    Paik, *The Vulgarity of Caste*, 62–63.
55. Geetha and Rajadurai, *Towards a Non-Brahman Millennium*, 64.
56. Ibid. Gail Omvedt, *Buddhism in India: Challenging Brahmanism and Caste* (New Delhi: Sage, 2003), 237.
57. Geetha and Rajadurai, *Towards a Non-Brahman Millennium*, 64.
58. Iyothee Thass, *Oru Paisa Tamizhan,* September 25, 1907, as cited in Geetha and Rajadurai, *Towards a Non-Brahman Millennium*, 62–63.
59. Gail Omvedt, *Dalits and the Democratic Revolution* (Delhi: Sage, 1994), 92.
60. Ibid., 92
61. Paik, *Dalit Women's Education in Modern India*, 83. In formal politics, Christophe Jaffrelot describes how the conservative character of the Congress Party promoted an "anti-casteist" rhetoric that corresponded with Gandhi's contradictory position on caste; Gandhi disproportionately shaped Congress' relatively conservative view of anti-caste politics—he supported the *varna* caste system for most of his life and opposed Ambedkar's advocacy for a separate electorate for Dalits. Jaffrelot, *India's Silent Revolution*.
62. "Constituent Assembly Debates, Official Report" (New Delhi: Lok Sabha Secretariat, 1949).
63. Ambedkar, "States and Minorities," 414.
64. Satish Deshpande, "Caste and Castelessness: Towards a Biography of the General Category," *Economic and Political Weekly* 48, no. 15 (2013): 32–39.
65. Ibid. Ajantha Subramanian, *The Caste of Merit* (Cambridge, MA: Harvard University Press, 2019); Ajantha Subramanian, "Making Merit: The Indian

Institutes of Technology and the Social Life of Caste," *Comparative Studies in Society and History* 57, no. 2 (2015): 291–32.
66. Subramanian, *The Caste of Merit.*
67. S. Deshpande, "Caste and Castelessness," 37.
68. Dorothy Roberts, "Sources of Commitment to Social Justice," *Roger Williams University Law Review* 4, no. 1 (1998): 175–203. Dalit, decolonial, anti-racist, and feminist scholars have shown how ideologies of liberalism, freedom, and liberty that undergird nation-building projects have consistently silenced marginalized groups.
69. Aloysius argues that using the term "Hinduism" itself is problematic as scholarship has pointed to a "colonial, collusive construct by power groups, native and foreign, referring mainly to the Brahmanical form." Aloysius, *Nationalism without a Nation in India*, 2.
70. Hindutva finds its origins in socio-religious reform movements in the nineteenth century that critiqued some aspects of caste but did not challenge the *varna* system. While early social reform movements like the Brahmo Samaj and Arya Samaj sought to construct a modern Hinduism, they failed to offer an anti-caste critique that challenged the Brahmanical ideology, as Jotirao Phule's Satya Shodhak Samaj did during the same period. Christophe Jaffrelot, "Introduction," in *Hindu Nationalism: A Reader*, ed. Christophe Jaffrelot, 3–25 (Delhi: Permanent Black, 2007).
71. In prison at the time for his participation in a plot to assassinate an associate of the Secretary of State in London, Savarkar's opinion shifted, and he came to believe that Muslims were the true enemies of the nation. For Savarkar, who identified as an atheist, Hindutva is not specific to "any particular creed or religious section thereof." Instead, he argues that "all those on this side of the Indus who claimed the land from Sindhu to Sindhu from Indus to the seas, as the land of their birth, felt that they were directly mentioned by that one single expression, Hindusthan." Vinayak Damodar Savarkar, "Hindutva: Who Is a Hindu," in *Hindu Nationalism: A Reader*, ed. Christophe Jaffrelot, 87–96 (Delhi: Permanent Black, 2007), 93.
72. Jaffrelot, "Introduction," in *Hindu Nationalism: A Reader*, 15.
73. Unlike most nationalist movements, Jaffrelot explains, "the RSS did not look at the conquest of state power as its priority." They instead sought to create a more "martial brand of Hinduism" and construct a dominant Hindutva identity by educating the population through organizing work in towns and villages where young men could meet every morning and evening for games and training with physical and ideological components. Christophe Jaffrelot, "Introduction," in *The Sangh Parivar*, ed. Christophe Jaffrelot, 1–22 (New Delhi: Oxford University Press, 2005), 2–3.
74. Muhammad Ali Jinnah was an important political leader of the Congress Party in the first two decades of the twentieth century. He joined the All-India Muslim League in 1913. Jinnah's secular views had decreasing

resonance and influence within the Congress Party after 1920, as Gandhi and his Hindu revivalism rose in prominence.

75. Jaffrelot, "Introduction," in *Hindu Nationalism: A Reader.*
76. S. Deshpande, "Caste and Castelessness," 35. Zelliot, *Ambedkar's World.*
77. Omvedt, *Buddhism in India.*
78. Shepherd, *God as Political Philosopher*, 19.
79. Ibid.
80. The RSS wanted to focus on expanding its network of *shakha*s throughout north India and beyond.
81. Jaffrelot, "Introduction," in *The Sangh Parivar*, 4
82. Shyama Prasad Mukherjee became the head of the Jana Sangh initially, but following his death in prison and after RSS activists forced out Mauli Chandra Sharma, who was elected to succeed Mukherjee, the Jana Sangh came under the leadership of RSS leader Deendayal Upadhyaya in 1954. Deendayal Upadhyaya and several other RSS leaders helped build the Jana Sangh into a grassroots organization that mirrored the RSS. The political party believed that religious minorities should "adopt Hindu cultural features and assimilate into a 'Hindian' nation," and its two well-known campaigns in the 1950s and 1960s involved advocating for Hindi to become India's national language and banning the slaughter of cows. Formal political power first emerged at the state level in the late 1960s through coalition governments in the Hindi-speaking heartland; the Jana Sangh formed coalition governments in Madhya Pradesh, Bihar, and Uttar Pradesh with partnerships with socialists and centrist parties following assembly elections in 1967, and thus had to pause its more radical agenda. Jaffrelot, "Introduction," in *Hindu Nationalism: A Reader*, 18.
83. A more moderate stream within the Jana Sangh presented itself in a populist vein of patriotism, pro-poor, and in support of small businesses; it guided the Jana Sangh's merger with other parties to form the Janata Party in 1977 to defeat Indira Gandhi and form the first non-Congress-led coalition government between 1977 and 1980. During this period, prime minister Morarji Desai appointed a Backward Classes Commission chaired by B. P. Mandal, which sought to extend reservations to OBCs. In 1980, the Janata Party leadership banned members from being dual members with the RSS (after RSS activist involvement in numerous Hindu–Muslim riots), which caused former Jana Sangh members with RSS membership to leave and create the BJP, with Atal Behari Vajpayee as its first leader. Jaffrelot, "Introduction," in *Hindu Nationalism: A Reader.*
84. Whether by converting to Buddhism, Islam, or Christianity or belonging to distinct local religious or spiritual communities that resisted processes of Hinduization, caste-oppressed groups have challenged majoritarian projects that justify caste hierarchy and untouchability. At the same time, as Kancha Illaiah Shepherd argues in *God as Political Philosopher* (2019: 23),

Brahmans have historically tried to dismantle radical challenges to Brahmanism (for example, Buddha, Basaveshwara, Ravidas) by "the deification of every person who opposed the caste system." Also see Joel Lee, *Deceptive Majority: Dalits, Hinduism, and Underground Religion* (Cambridge: Cambridge University Press, 2021); Omvedt, *Buddhism in India*; Geetha and Rajadurai, *Towards a Non-Brahman Millennium*; Mark Juergenseyer, *Religion as Social Vision* (Berkeley: University of California Press, 1982); B. R. Ambedkar, "The Buddha and His Dhamma," in *Dr. Babasaheb Ambedkar: Writings and Speeches, Vol. 11*, ed. Vasant Moon, 9–599 (New Delhi: Dr. Ambedkar Foundation, 1979); Jotirao Phule, "*From* The Book of the True Faith (1891)," in *Selected Writings of Jotirao Phule*, ed. G.P. Deshpande, 224–36 (New Delhi: Leftworld Books, 2002); Jotirao Phule, "Cultivator's Whipcord (1883)," in *Selected Writings of Jotirao Phule*, ed. G.P. Deshpande, 114–89 (New Delhi: Leftworld Books, 2002).

85. Jean Drèze, "The Revolt of the Upper Castes," *CASTE/A Global Journal on Social Exclusion* 1, no. 1 (2020): 229–36.
86. Guru, "Liberal Democracy in India and the Dalit Critique."
87. As India's first law minister, Ambedkar tried to pass a Hindu Code Bill that challenged caste and patriarchy in social life, but political leaders blocked his efforts, and he resigned as law minister citing internal resistance to change.
88. In their critique of decolonizing institutions in the 1970s, the Dalit Panthers described how the state has supported the monopoly ownership of land, industry, and economic wealth, and allowed for religiously sanctioned caste-based domination to continue. The Panthers critiqued Brahmanical control of the state apparatus and the interconnections of a political and economic system that simultaneously (re)produced caste power and capitalist growth in a manner that concentrated the benefits of public and private resources for caste elites. Dalit Panthers, *Manifesto* (Bombay, 1973).
89. S. Deshpande, "Caste and Castelessness," 36–37.
90. Ibid., 36–37.
91. Galanter, *Competing Equalities*.
92. S. Deshpande, "Caste and Castelessness," 36–37.
93. Subramanian, "Making Merit."
94. Aloysius, *Nationalism without a Nation in India*. Caste-oppressed groups fought to create public spaces which had been restricted by Brahmanism—whether arguing for publicly funded roads, rails, and wells available for all or against separate schools for children considered "untouchable." V. Geetha and S. V. Rajadurai document how Adi-Dravidas ("untouchable") and non-Brahman groups in Madras "facilitated an expansion, literally and ideologically, of civil society" through protests over the use of space, as well through alternative presses and dailies in the second half of the nineteenth and early twentieth centuries. Geetha and Rajadurai, *Towards a Non-Brahman Millennium*, 60.

95. Suryakant Waghmore, *Civility against Caste* (Delhi: Sage, 2013), 201.
96. Benedict Anderson, *Imagined Communities: Reflections on the Origin and Spread of Nationalism.* (London: Verso, 1983); Bruce Curtis, *The Politics of Population: State Formation, Statistics and the Census of Canada 1840–1875* (Toronto: University of Toronto Press, 2001); David I. Kertzer and Dominique Arel, "Censuses, Identity Formation, and the Struggle for Political Power," in *Census and Identity: The Politics of Race, Ethnicity, and Language in National Censuses*, ed. David I. Kertzer and Dominique Arel, 1–42 (Cambridge: Cambridge University Press, 2002).
97. Tianna S. Paschel, *Becoming Black Political Subjects* (Princeton: Princeton University Press, 2016), 133.
98. Wendy Nelson Espeland and Mitchell L. Stevens, "A Sociology of Quantification," *European Journal of Sociology* 49, no. 3 (2008): 401–36; Mara Loveman, "The Modern State and the Primitive Accumulation of Symbolic Power," *American Journal of Sociology* 110, no. 6 (2005): 1651–83; Alain Desrosières, *The Politics of Large Numbers: A History of Statistical Reasoning* (Cambridge, MA: Harvard University Press, 1998); James Scott, *Seeing Like a State* (New Haven: Yale University Press, 1998); Theodore M. Porter, *The Rise of Statistical Thinking, 1820–1900* (Princeton: Princeton University Press, 1986); Benedict Anderson, *Imagined Communities: Reflections on the Origin and Spread of Nationalism* (London: Verso, 1983).
99. Pierre Bourdieu, *Language and Symbolic Power*, ed. John Thompson (Cambridge: Polity Press, 1991).
100. James Scott, *Seeing Like a State* (New Haven: Yale University Press, 1998).
101. Curtis, *The Politics of Population.*
102. Nicholas Dirks, *Castes of Mind: Colonialism and the Making of Modern India* (Princeton, NJ: Princeton University Press, 2001); Bernard Cohn, "The Census, Social Structure and Objectification in South Asia," in *An Anthropologist among the Historians and Other Essays*, 224–54 (Delhi: Oxford University Press, 1987); Arjun Appadurai, "Number in the Colonial Imagination," in *Orientalism and the Postcolonial Predicament: Perspectives on South Asia*, ed. Carol A. Breckenridge and Peter van der Veer, South Asia Seminar Series, 314–36 (Philadelphia: University of Pennsylvania Press, 1993).
103. Melissa Nobles, *Shades of Citizenship: Race and the Census in Modern Politics* (Stanford, CA: Stanford University Press, 2000).
104. Michael Rodríguez-Muñiz, "Cultivating Consent: Nonstate Leaders and the Orchestration of State Legibility," *American Journal of Sociology* 123, no. 2 (September 2017): 385–425. Rodríguez-Muñiz builds on works by Bruce Curtis and Geoffrey Bower in defining census infrastructure work. Curtis, *The Politics of Population.* Geoffrey C. Bowker, *Science on the Run: Information Management and Industrial Geophysics at Schlumberger, 1920–1940* (Cambridge, MA: MIT Press, 1994).

105. Rodríguez-Muñiz, "Cultivating Consent."
106. Mara Loveman and Jerônimo O. Muniz, "How Puerto Rico Became White: Boundary Dynamics and Intercensus Racial Reclassification," *American Sociological Review* 72, no. 6 (2007): 915–39; Mara Loveman, "The U.S. Census and the Contested Rules of Racial Classification in Early Twentieth-Century Puerto Rico," *Caribbean Studies* 35, no. 2 (2007): 79–114. Sociologists Aliya Saperstein and Andrew Penner also find that survey interviewers' classifications of race vary over time in relation to their perceptions of the respondents' circumstances. Their research speaks both to the fluidity of racial categorization by enumerators and the social embeddedness of racial data. Aliya Saperstein and Andrew Penner, "Racial Fluidity and Inequality in the United States," *American Journal of Sociology* 118, no. 3 (2012): 676–727; Aliya Saperstein and Andrew Penner, "Beyond the Looking Glass: Exploring Fluidity in Racial Self-Identification and Interviewer Classification," *Sociological Perspectives* 57, no. 2 (2014): 186–207.
107. Wendy Nelson Espeland and Mitchell L. Stevens, "Commensuration as a Social Process," *Annual Review of Sociology* 24 (1998): 313–43, 314.
108. Ibid., 328.
109. Ibid., 314, 316.
110. Ibid., 317.
111. Ibid.
112. Ibid.
113. Ibid.
114. Bruno Latour, *Science in Action* (Cambridge, MA: Harvard University Press, 1987).
115. Palashi Vaghela, Steven J. Jackson, and Phoebe Sengers, "Interrupting Merit, Subverting Legibility: Navigating Caste in 'Casteless' Worlds of Computing," in *Proceedings of the 2022 CHI Conference on Human Factors in Computing Systems (CHI "22)*, Article 545 (New York, NY: Association for Computing Machinery, 2022), 1–20; the authors (2022: 13) highlight the modernity of caste and in doing so how caste is neither residual nor limited to social life but a "force within the practice of computing workplaces themselves." For more on critical caste and technology studies, see Murali Shanmugavelan, "May the Myth of Castelessness Die," *LOGIC(S)*, no. 19 (2023).
116. Curtis, *The Politics of Population*, 4.
117. Appadurai, "Number in the Colonial Imagination."
118. Mara Loveman, *National Colors: Racial Classification and the State in Latin America* (Oxford: Oxford University Press, 2014).
119. Ibid., 85, 132.
120. Nobles, *Shades of Citizenship*; Curtis, *The Politics of Population*; Paschel, *Becoming Black Political Subjects*; Debra Thompson, *The Schematic State:*

Race, Transnationalism, and the Politics of the Census (Cambridge: Cambridge University Press, 2016).

121. Loveman, *National Colors*, 12.
122. Alain Blum, "Resistance to Identity Categorization in France," in *Census and Identity: The Politics of Race, Ethnicity, and Language in National Censuses*, ed. David I. Kertzer and Dominique Arel, 121–47 (Cambridge: Cambridge University Press, 2002).
123. Interview with activist, February 22, 2013. The activist explained how "only the ORGI has the technical skill" to collect caste-wise data but that "the bureaucracy had never seen it in their interest to collect the data because it remains mostly 'upper' caste." In addition, he discussed the complexity of OBC politics in which groups that had disproportionately benefited from OBC reservations also did not want a caste census.
124. Ambedkar, "Thoughts on Linguistic States," 167.
125. Dirks, *Castes of Mind.*
126. Cohn, "The Census, Social Structure and Objectification in South Asia," 242–43.
127. For example, Jotirao Phule dedicated his 1873 book *Slavery: In the Civilized British Government under the Cloak of Brahmanism* to the "good people of America ... that my country men may take their noble example as their guide in the emancipation of their Sudra Brethern from the trammels of Brahman thralldom." His scathing critique of Brahmanism in the late nineteenth century documents the centrality of caste to the expansion of colonial institutions in India. Phule, "Slavery (1873)," 25.
128. B. R. Ambedkar and W. E. B. Du Bois were both scholar-activists committed to challenging systems of caste and exploitation in their homelands and abroad. In a letter that Ambedkar wrote to Du Bois in 1946, he highlights the connections between caste and race and a comparative view of systems of oppression in the twentieth century:

    > Although I have not met you personally, I know you by name as everyone does who is working in the cause of securing liberty to the oppressed people. I belong to the Untouchables of India and perhaps you might have heard my name. I have been a student of the Negro problem and have read your writings throughout. There is so much similarity between the position of the Untouchables in India and of the position of the Negros in America that the study of the latter is not only natural but necessary.

    Ambedkar saw a scholarly examination and understanding of racism in the United States as both "natural" and "necessary" in strengthening his efforts to dismantle untouchability and caste in India. B. R. Ambedkar, "Letter from B.R. Ambedkar to W.E.B. Du Bois," Undated (likely written in 1946), South Asian American Digital Archive, https://www.saada.org/item/20140415-3544, accessed April 1, 2024.

2
The Institutional Life of Caste

> Let this Committee and let the whole world know that today there is a body of Hindu reformers who are pledged to remove this blot of untouchability. We do not want on our register and on our census untouchables classified as a separate class.
>
> —M. K. Gandhi, 1931[1]

M. K. Gandhi opposed the census enumeration of "untouchables," or Dalits, as a separate group in the Census of 1931. As a "Hindu reformer" who fought against untouchability, Gandhi wanted to include "untouchables" within the larger political category of "Hindu." Yet this view ran against the lived experiences of Dalits, who pointed to the hypocrisy of the political construction of "untouchables as Hindus" when they were systematically excluded, humiliated, and treated as less than human, and blocked from entry into temples—reinforcing the "line of untouchability" that separated "untouchables" from "caste Hindus." Political scientist Vivek Kumar Singh describes the irreconcilability of this position, arguing that "untouchables could, therefore, neither enter the temple nor leave it."[2]

Ambedkar vehemently opposed Gandhi's position. Gandhi saw himself as representing "the vast mass of untouchables" and believed that "untouchables" should remain within the Hindu fold, while Ambedkar demanded self-representation for "untouchables" through separate electorates and reservations.[3] Gandhi resisted both separate electorates and reservations and also argued that the census enumeration of "untouchables" as a separate community would further enhance caste divisions, instead of empowering "untouchables." Gandhi's perspective in the 1930s strongly influenced Congress political leaders, who believed they were the legitimate political representatives of "untouchables" as one of many communities within the "Hindu majority" during negotiations with the British to "quit India."

Gandhi's advice to a government committee on how "untouchables" should be enumerated in the census—as Hindus—highlights the embeddedness of census politics within the politics of representation and the distribution of political power in "decolonizing" India.[4] Gandhi's recommendation also reveals the gap between lived experiences of caste oppression and the emerging nationalist constructions of a casteless Hindu citizen, which in practice was inaccessible for "untouchables," other caste-oppressed groups, and Muslims. Congress's emerging conception of the "Indian citizen" sought to bring "untouchables" within the political fold of a "Hindu majority" without supporting movements to destroy caste hierarchy and dismantle caste privilege—thereby rejecting the approach of anti-caste movements fighting to demolish systems of internal and external colonization. Debates about the political representation and categorization of "untouchables" and religious minorities were deeply intertwined with the construction of an Indian national identity, processes of enumeration in the census, the ability for enumerated groups to make claims on the state, and the fight for political power.

This chapter traces the changing relationship between caste hierarchy and the census. I first examine caste in the colonial census. I then discuss the embeddedness of Brahmanism within the colonial state—initially in ways that made caste hierarchy visible and later through an ideology of castelessness.[5] Finally, I explore key changes that occurred in the decennial censuses of independent India, after political leaders removed a caste-wise enumeration from the census.

Caste and Census in the Colonial Period

The colonial obsession with classification by caste and religion was intertwined with the broader project of empire building.[6] Along with the desire to politically control territory and exploit resources for economic benefit, colonialism was "a cultural project of control" undergirded by beliefs of racial, religious, and cultural superiority.[7] Colonial knowledge "both enabled conquest and was produced by it; in certain important ways, knowledge was what colonialism was all about."[8] As part of early survey mappings of the subcontinent, British colonial administrators found that the simplest enumerations of local populations were challenging given the administrators' ignorance of social relations and difficulties in translating social relations into discrete categories.[9]

In response to armed resistance against British rule between 1857 and 1859, the colonial government desired to develop more comprehensive and systematic knowledge about its subjects.[10] Anthropological historian Bernard Cohn traces how among colonizers "there was widespread agreement that this society, like others they were governing, could be known and represented as a series of facts," and that the facts were "taken to be self-evident, as was the idea 'that administrative power stemmed from the efficient use of these facts.'"[11] Along with learning local languages, colonial administrators believed that knowledge about religion and caste was the sociological key to understanding the colonized, and data on caste and religion fed into army recruitment, balancing the proportions of Hindus and Muslims in public service, developing education policy, and monitoring organized efforts to end British rule.[12] As part of the broader "enumerative modality" that began with British merchants in the early seventeenth century, the colonial state started to centralize an extensive census apparatus in the second half of the nineteenth century to produce tables and monographs about caste.[13] Administrators, such as W. W. Hunter, the first director-general of statistics and later director general of gazetteers, felt that precise data was key to the successful rule of the subcontinent.[14] Decennial censuses beginning in the second half of the nineteenth century mobilized extensive resources, including 500,000 enumerators, for data collection across the territory.[15]

Imperial censuses in the subcontinent were far more detailed in scope than the censuses being carried out in Britain during the same period. Census-makers in Britain did not collect data on religion or race until well after the end of the British rule in India, while in colonial censuses religion, caste, and race emerged as fundamental categories of enumeration.[16] The Crown carried out many enumerative practices in British India prior to implementing them at home. Bernard Cohn explains:

> It is not just that the personnel who governed India were British but the projects of state building in both countries—documentation, legitimation, classification, and bounding, and the institutions therewith—often reflected theories, experiences, and practices worked out originally in India and then applied in Great Britain, as well as vice versa.[17]

In nine decennial censuses between 1871 and 1941, the colonial administration set up individual census commissions to classify the entire population within the territory by religion and caste. In the 1872 Census report, more than a fourth of the content focused on religion and caste data.[18] In the 1881 Census, the proliferation of caste names in the collected data led to the simultaneous publication of "caste index" volumes, along with the main census report.[19]

In the 1901 Census, a quarter of the main report focused on caste and religion data, and the publication of caste appendices and supplemental reports were much greater in length than the main census report. The final enumeration of caste in 1941 remained un-tabulated and unpublished due to the hasty collection and compilation of data against the backdrop of World War II.

The collecting and compiling of caste and religion data consisted of processes of *commensuration*, involving census officials, enumerators, community leaders, and households. Commensuration sought to put into a common metric differing conceptions of caste and religion; it entailed creating a system "for discarding information and organizing what remain[ed] into new forms."[20] Census administrators' limited understanding of social relations—and the complexity of documenting local caste hierarchies and creating comparable categories across local and regional hierarchies—led to a messy process of commensurating caste and religion within and across imperial censuses (see Table 2.1). As early as 1881, regional census officers started to develop caste lists prior to the official census enumeration to aid in commensuration; these lists included a standardized name for castes and tribes and the variations of the name found throughout the region.[21] Supervisors, usually a government employee or another literate member of the community, instructed enumerators how to classify responses, including advice on terms to avoid because they were too "generic" or otherwise differed from the desired data.[22] By 1891, all regional census directors had compiled lists of castes and subcastes and drafted special instructions for the enumeration of caste in their region. Commensuration occurred at every stage of the process—during the creation of caste lists before the start of the census, throughout enumeration, and in post-enumeration tabulation. Census officials devoted considerable time to finalize the caste tables despite the creation of regional caste lists. The census commissioner of 1891 estimated that compiling caste data took 25 percent of the tabulation time for the census.[23]

Elite Indians played an important role in the production of caste data and helped to create a particular social map bolstered by references to religious texts. In the 1881 Census, Rajendra Lal Mitra, a leading Sanskrit scholar at the time, determined the relative social ranking of castes in Bengal based on "textbooks of the Hindus" and objected to census officials directly adjudicating claims by groups for higher positions in the caste hierarchy by stating: "[the census'] duty is clearly to follow the textbooks of the Hindus and not to decide on particular claims."[24] In the 1901 Census, census commissioner H. H. Risley relied almost entirely upon the opinions of Brahmans and other

Table 2.1 Caste in the Colonial Census (1871–1941)

Year	Census questions	Notes
1871/ 1872	Q5/12: Religion Q6/13: Caste or Class Q7/14: Race or Nationality	• Non-simultaneous census (time and duration of enumeration determined by local authorities between Nov. 1871 and Mar. 1872) yet first systematic enumeration of population of British India. • Head of household empowered to give or withhold names of female members. As such, separate questions for enumeration of religion, caste, and race of males (Q5, 6, 7) from enumeration of females (Q12, 13, 14).[1] • Recognized need to enumerate religion and caste separately for each individual in household (draft of questionnaire included combined column for "religion or caste" and later separated). An assistant commissioner of Oudh pointed out that without separate columns for religion and caste "all the lower castes of Mussulmans, e.g. Joolahs, Kassaees, Ghosees, Durzees, will inevitably be entered as Mussulamans without any detail of caste."[2]
1881	Q6: Religion Q7: Caste, if Hindu; sect, if of other religion	• First synchronous census. • Local authorities in some regions drew up relevant list of castes. Madras Province drafted special rule to instruct enumeration of caste and sect.[3] • Caste was associated with Hindus and sect with all other religions; confusion resulted. Census officials found question on sect most relevant for Christians.
1891	Q2: Religion Q3: Sect of Religion Q4: Caste of Hindus & Jains. Tribe or race of others Q5: Sub-Div. of caste &c.	• Each province drafted special instructions for the enumeration of caste. • Census commissioner estimated that compiling caste data took 25% of tabulation time.[4] • Enumerator instructions for Q2 specify to enter religion for each person "as Hindu, Mussulman, Jain, Christian, Parsi. Forest tribes, who are not Hindus, Mussulmans, &c., should have the names of their tribe entered in this column, as Bhil, Gound, Garo, &c., low castes as Chamar, Dom, Paria, Mhar, &c., should be entered by the religion which they themselves return, and no dispute about it is to be raised."

(*Continued*)

Table 2.1 (*Continued*)

Year	Census questions	Notes
		• Enumerator instructions for Q3 provide examples of sect for Hindus, Muslims, and Christians and specify to write "not returned" if sect not shared and not to leave answer blank. • The wording of Q4 implies that caste is associated with Hindu and Jain religions. However, enumerator instructions provide examples of caste for Hindus and Muslims, as well as examples of race for Christians (i.e., Eurasian or Native). Instructions specify that regional identities (e.g., Hindustani, Marwari, Punjabi) are "vague" and should not be entered.[5] • Enumerator instructions for Q5 provide examples of subcastes for Hindus (e.g., the subdivision of Brahmin), specification of clan for tribes, and details on tribe or nationality for race. For native Christians, caste could be entered for Q5. If castes or races do not have subdivision, information entered for Q4 may be repeated from Q5, but Q5 should not be left blank.
1901	Q4: Religion Q8: Caste of Hindus & Jains. Tribe or race of others	• Enumerator instructions for Q4 specify "enter here the religion which each person returns, as Hindu, Mussulman, Jain, Christian, Parsi. In the case of Christians the sect also should be entered below the religion." • Related to Q8, enumerator instructions specify to "enter the caste of Hindus and Jains, the tribe of those who have no castes, and the race of Christians, Buddhists, etc."[6] • Supervisors provided enumerators with additional written instructions for the enumeration of caste and religion based on instructions drafted at the provincial level. • Published report included first hierarchical ranking of castes.
1911	Q4: Religion Q8: Caste, Tribe, or Race	• Enumerator instructions related to Q4 specify that people should be enumerated as "Hindu, Musalman, Sikh, Jain, Christian, Parsi. In the case of Christians the sect should also be entered. In the case of aboriginal tribes who are not Hindus, Musalmans, Christians, Etc., the name of tribe should be entered in this column."

(*Continued*)

Table 2.1 (*Continued*)

Year	Census questions	Notes
		• Related to Q8, enumerator instructions specify to "enter the caste or tribe of Hindus, Musalmans, Jains, Sikhs, Aryas, Brahmos and aboriginal tribes, and the race of Christians, Buddhists, Parsis, etc."[7]
1921	Q4: Religion Q8: Caste, Tribe or Race	• For Q4 and Q8, same enumerator instructions as 1911 census. • Category of "depressed classes" used for first-time as all-India category of caste oppressed groups.
1931	Q4: Religion and sect Q8: Race, Tribe or Caste	• Enumerator instructions related to Q4 instruct "enter here the religion which each person returns, as Hindu, Muslim, Sikh, Jain, Christian, Parsi and the sect where necessary. In the case of aboriginal tribes who are not Hindu, Buddhist, Christian, etc., the name of the tribe should be entered in this column. Sect is in all cases required for Christians." • Enumerator instructions for Q8 explain that "for Indians enter caste as ordinarily understood but for wide castes enter sub-caste also. The class titles-Brahman, Kshatriya, Vaishya and Sudra are usually insufficient by themselves. For other subjects of the Empire and for foreigners enter race as 'Anglo-Indian', 'Canadian', 'Goanese', 'Turkish.' For Indians such as some Christians who have no either caste nor tribe, enter 'Indian'."[8] • During tabulation of caste data, "exterior groups" used as an all-India category for oppressed groups.
1941	Q3: Race, Tribe or Caste Q4: Religion	• Caste and religion data not tabulated or published due to war-time financial and resource limitations. • Enumerator instructions conceive of caste or tribe (Q3) as categories associated with all Indians (across religious groups) while "in the case of non-Indians, enter the race or nationality."

(*Continued*)

Table 2.1 (*Continued*)

Year	Census questions	Notes
		• Enumerator instructions for Q4 include not only previously listed religions but also "Agnostic" and "Confucian". They also specify "if tribal name given record it."[9]

Source: Dandapani Natarajan, *Indian Census through a Hundred Years*, Census of India 1971 (Delhi: Office of the Registrar General & Census Commissioner, India, 1972).

Notes:

[1] Dandapani Natarajan, *Indian Census through a Hundred Years*, Census of India 1971 (Delhi: Office of the Registrar General & Census Commissioner, India, 1972), 356.

[2] Ibid., 186.

[3] J. A. Baines, "Conference of Census Officers of the 1881 Census to Consider the Matter relating to 1891 Census (Feb 10th, 1890)," in ibid., 246.

[4] J. A. Baines, "Letter to George Hamilton, dated April 29, 1899," in ibid., 249.

[5] Ibid., 537–38.

[6] Ibid., 550.

[7] Ibid., 567.

[8] Ibid., 606.

[9] Ibid., 625.

"*higher* castes" to develop a ranking of castes. Anthropological historian Nicholas Dirks argues that the Brahmanical perspective of other castes—particularly with regard to from whom a person may accept food and water, the ritual proximity of groups in relation to Brahmans, and origin stories concerning duties and obligations of other castes toward Brahmans—became the basis for the official census ranking of castes.[25] Caste elites helped to produce and rank caste categories, and their respective social maps disproportionately shaped the creation of colonial state knowledge. Colonial administrators and caste elites who served as experts in analyzing caste data constructed what Dirks calls a "Brahmanical sociology."[26]

Groups mobilizing against untouchability and caste hierarchy, as well as those who were reformist in their orientation, also organized in relation to the census enumeration of caste. Iyothee Thass, who fought against caste-based oppression in the Madras Presidency in the late nineteenth century, founded the Advaidananda Sabha in 1870 in the Nilgiris to fight caste hierarchy and proselytization of Christian missionaries; he later urged Dalits to embrace Buddhism.[27] In 1881, after census officials classified Dalits within the Hindu religion, Thass wrote a memorandum to the colonial government stating that oppressed castes in Tamil-speaking areas should be classified as "Adi-Tamizhar" (the original Tamils) and not Hindu, but the government did not agree to this change. In the lead-up to the Census of 1891, he urged Dalits to register themselves not as Hindus but as "casteless Dravidians."[28] The census became one of multiple sites to make political claims against caste apartheid and challenge the naming and treatment of oppressed castes. In 1891, Thass co-created the Dravida Mahajana Sabha, and its members passed ten resolutions at their first conference, including a resolution to enact a criminal law to punish those who humiliated caste-oppressed groups by calling Dalits "pariahs." As such, the politics of enumerating caste in colonial censuses intersected with broader anti-caste organizing efforts during this period. First, it helped Dalits and other caste-oppressed groups fight for self-representation as political subjects and make claims on the state. As historian Pritam Singh explains, "the colonial census formed a critical instrument that could help in understanding the marginalization of the lower castes."[29] Census data were used to justify oppressed castes' claims for self-representation during negotiations for political reform during the Montagu–Chelmsford Reforms of 1919, the Simon Commission of 1928, and the Round Table Conferences in the early 1930s.[30] Leaders of caste-oppressed groups who secured "a seat at the table" then used census data to negotiate for the development of new policies to fight against untouchability and caste hierarchy. They also fought for access to existing government programs and institutions—from education to employment—some of which they had formal access to but were excluded from in practice. In addition, anti-caste leaders mobilized caste-oppressed communities to counter dominant narratives by developing new subjectivities and self-understandings and reclaiming histories that centered the contributions of caste-oppressed groups—and utilized censuses in these efforts. For example, they refused to be counted in the Hindu fold—which deemed them "outcastes," "untouchables," and "pariahs"—when mainstream nationalists sought their numbers to construct a Hindu majority but failed to challenge caste hierarchy and untouchability. These movements challenged ideologies of Brahmanism and Hindutva and the political legitimacy of elites and reformist organizations that spoke out against British exploitation but remained complicit with internal systems of caste domination. Caste-oppressed groups that mobilized to counter the mainstream nationalist movement took considerable heat from political elites who desired to be all-India representatives, as described in the opening vignette.

Other movements during this period also organized in response to the census but were reformist in their orientation and did not seek to dismantle caste hierarchy. The Arya Samaj, a religious movement aimed at developing a more revived and purified Vedic religion in northwestern India, came to reject the label of Hindu.[31] In the two years leading up to the 1891 Census, Arya Samaj newspapers instructed their followers to answer "Arya" instead of "Hindu" for the religion question. In addition, they also pressured census officials to accept and record "Arya" as a religion.[32] Forty years later, the Arya Samaj continued its mobilization in the lead up to the 1931 Census—this time urging its members to specify that they had no caste during the census.[33] The Arya Samaj widely distributed a handbill in Lahore that instructed how to answer census questions (Figure 2.1).

A proliferation of caste associations emerged in the aftermath of the census ranking of castes in 1901.[34] Many of them tried to shape the production of caste tables by lobbying census officials. The census office became a key site of mobilization for the contesting of a group's relative position in the caste hierarchy. While numerous caste associations fought for changes in their own caste's relative ranking, they less commonly sought to destroy caste hierarchy and untouchability as Thass and other caste-oppressed leaders fought to do.

The census involved extensive processes of commensuration that worked to standardize and aggregate caste and religion data both within and across regions. As Bernard Cohn argues, "from the beginning of the census operations it was widely assumed that an all-India system of classification of castes could be developed."[35] In the 1872, 1881, and 1901 Censuses, the colonial state used the *varna* system—the fivefold categorization of Shudras, Vaishyas, Kshatriyas, Brahmans, and groups considered outside the caste system such as "Untouchables" (that is, Dalits) or "Aborigines" (that is, Adivasis)—to classify the population.[36] While caste roughly mapped onto *varna* in certain regions of north India, the system had limited application in much of the northeast, east, and south. In many regions, the caste and religious composition made

Remember!
Census Operations Have Begun

Question	You should answer!
Religion	Vedi Dharm
Sect	Arya Samajist
Caste	Nil
Race	Aryan
Language	Arya Bhasha

Figure 2.1 Arya Samaj Handbill in Lahore (1931)

Source: India Census Commissioner, Census of India, Vol. XVII, 1931.

little sense within the *varna* framework. In the 1881 Census, commissioner W. C. Plowden wanted to create caste tables that included data on all castes with more than 100,000 people and utilized the *varna* categories despite protests from regional census commissioners.[37] Plowden then abandoned *varna* categories in the next census, given the disconnect between the *varna* system and caste hierarchy in many regions. However, when H. H. Risley took over control of the census in 1901, he returned to the *varna* system. The re-adoption of the *varna* system continued the messy process of constructing all-India caste categories.[38] While previously individuals from one region of the country had limited understanding of caste names and relationships in other regions, the colonial state's adoption of top-down, all-India systems of classification contributed to people beginning to conceptualize themselves in relations to others beyond localized caste hierarchies. Yet problems with this metric of commensuration abounded and, in the compilation and publication of caste data in the 1901 Census, Risley abandoned a pan-Indian system of classification and included a ranking of caste hierarchy for the first time.[39] The 1901 Census became "the official record of social status" and, as already discussed, caste associations began to lobby and argue for changes in the relative ranking of castes in the census results and filed petitions for higher status.[40] Depressed classes emerged as the official term for the most socially and economically disadvantaged groups, and census returns for the 1921 Census reported the population of this all-India category.[41] Legal scholar Marc Galanter discusses how the introduction of this broad term "opened the possibility of visualizing the problem not as that of a congeries of depressed groups, but as a stratum of all-India dimension with shared characteristics."[42] Caste tables in the census also provided data by religion and documented how within a particular caste, such as the "untouchable" Chamar caste, individuals identified as Hindu, Sikh, Arya, Muslim, Jain, Buddhist, and belonging to tribal religions.[43] The 1931 Census categorized oppressed groups as "exterior castes" and published the final caste-wise tables from a decennial census. Across successive censuses, census officials developed and tested national-level categories. These all-India categories strengthened and emerged in response to movements to dismantle caste hierarchy and untouchability and expand political representation, educational access, and pathways to government employment for caste-oppressed groups.

Census schedules, enumerator instructions, and published tables in census reports reveal a changing understanding of the relationship between caste and religion. In the first colonial census, the enumeration of caste included

Muslim communities—it was the stated justification for changing a combined question on religion and caste to two separate questions. An assistant commissioner of Oudh pointed out that without separate columns for religion and caste, "all the lower castes of Mussulmans [Muslims], e.g. Joolahs, Kassaees, Ghosees, Durzees, will inevitably be entered as Mussulamans without any detail of caste."[44] In the second colonial census, enumerators were instructed to enumerate caste for Hindus and sect for other religions.[45] However, this approach proved confusing and inadequate to record how caste affected religious minorities. In the 1891 and 1901 Censuses, the wording of the caste question implied that caste was only relevant for Hindus and Jains. The first part of the enumerator instructions conveys this view when it states, "[E]nter the caste of Hindus and Jains, and tribes of those who have not castes, and the races of Christians, Buddhists, etc."[46] The next three censuses envisioned caste or tribe as an identity that could be enumerated for Indians across most religious groups. In the 1911 and 1921 Censuses, for example, the enumerator instructions clearly direct, "[E]nter the caste or tribe of Hindus, Musalmans, Jains, Sikhs, Aryas, Brahmos and aboriginal tribes, and the race of Christians, Buddhists, Parsis, etc."[47] The minority communities of Christians, Buddhists, and Parsis remained associated with race and not caste or tribe. The final two colonial censuses recognized caste among all Indians. In the 1931 Census, the instructions for the question on caste specify, "[F]or Indians, enter caste as ordinarily understood."[48] Although the 1941 Census did not publish collected caste-wise data due to wartime constraints, the enumerator instructions similarly conceived of caste and tribe as categories relevant across religious groups. They state, "[F]or Indians enter the caste or tribe returned. In the case of non-Indians, enter the race or nationality."[49] Census classifications across the colonial period had changing conceptualizations of caste from something relevant to specific religious communities to a category relevant for all Indians across religious groups. The commensuration of caste during the colonial census involved standardizing caste names within a region, creating administrative categories that allowed for an all-India system of classification, and shifting the boundaries of caste in relation to religious identities.

A body of scholarly work has also asserted that colonial censuses produced caste. Those advocating against a caste-wise enumeration in the current day often reference this historical research. Scholars have argued that prior to colonial rule various forms of hierarchical social relations existed but that these systems differed from contemporary conceptualizations of caste. As colonial censuses struggled to classify social relationships across successive censuses, historian Padmanabh Samarendra asserts that the census created a new conceptual system of classification by applying the *varna* system unsuccessfully, and then instead capturing *jati* relations and trying to map them onto all-India administrative categories.[50] Samarendra states:

> The project entailed that the diverse jatis be first counted, and then classified within a new pan-Indian template. The obligation was novel and unparalleled. The jatis, along with their assumed numerical strength had been enlisted earlier too; however, these communities stood in such lists as discrete units. The summing up of number in the census, on the other hand, was possible only if the entities counted were made comparable. The compulsion, under the circumstances, was to find the features common to the otherwise divergent jatis, so that these could be defined and demarcated uniformly.[51]

Samarendra, as well as Bernard Cohn and Nicholas Dirks in their earlier research on colonial censuses, documents processes of commensuration during the production of social data. Bruce Curtis describes this process as creating "social equivalents" among individuals and groups, in his historical study of Canadian censuses.[52] The creation of "social equivalents" is a social and political activity shaped by the worldviews of census administrators, social norms and power relations, respondents' points-of-view, worldviews of data collectors and analyzers, and broader state objectives and nation-building goals. Classifying individuals or households by caste, race, religion, gender, or nationality is not a preset fact, but instead reflects a historically specific understanding of an identity, which may be actively contested even at the time of classification. While colonial censuses sought to commensurate caste as a comparable and aggregable category across individuals, households, regions, and groups, there is no evidence that colonialism created caste hierarchy, although like all transformative social, economic, cultural, and political processes it shaped understandings and experiences of caste hierarchy.

The massive effort to enumerate the population by caste and religion reflected the colonial state's attempt to make sense of and create governable subjects. Colonial censuses repeatedly tried to create national-level caste categories as a means of organizing the thousands of castes and subcastes that make up localized systems of stratification throughout the subcontinent. Leading sociologists and anthropologists of South Asia, including G. S. Ghurye, M. S. Srinivasan, Bernard Cohn, Arjun Appadurai, and Nicholas Dirks, have argued that the colonial obsession with classification by caste and religion was deeply intertwined with the English determination to "divide and rule."[53]

At the same time, as discussed previously, these national-level categories aided in the development and implementation of centralized policies to address historic and ongoing caste-based discrimination and exclusion. Census data on caste shaped the colonial state's social map of society and scholarly understandings of caste. Bernard Cohn argues in his 1984 essay, "The Census, Social Structure and Objectification in South Asia":

> Most of the basic treatises on the Indian caste system written during the period 1880 to 1950 were written by men who had important positions either as census commissioners for all of India or for a province.... It would not be an exaggeration to say that down until 1950 scholars' and scientists' view on the nature and structure and functioning of the Indian caste system were shaped mainly by the data and conceptions growing out of the census operations.[54]

While Cohn's research focused on South Asia and the role of censuses in the construction of colonial knowledge and power, beyond the region, Michel Foucault highlighted the nexus of knowledge production and state power, and Bourdieu pointed to the importance of symbolic power in the operation of contemporary states.[55] Cohn's attention to colonial censuses as a key knowledge-making project, as well as the effects of census-making and knowledge production in the construction of state power, led him to overstate that censuses and census commissioners were the key producers of knowledge about caste between 1880 and 1950. Powerful anti-caste movements seeking to eradicate caste and challenge Brahmanism emerged and strengthened during the second half of the nineteenth and early twentieth centuries. Leaders of these movements wrote scathing critiques of how oppressor castes controlled knowledge-making both within and outside the state and how caste-based power operated. Scholars contemporary to Cohn, including several trained in the United States like Cohn, wrote extensively in the 1970s and 1980s about the writings and activism of Ambedkar and Phule, as well as social, literary, and spiritual movements against untouchability and caste hierarchy, including the non-Brahman and Ad Dharm movements—offering a comprehensive analysis of how caste hierarchy operated.[56] I now turn to the primary writings of activist-scholars who critiqued caste hierarchy and the embeddedness of Brahmanism within institutions beginning in the final quarter of the nineteenth century and continuing into the period of decolonization from England. I also draw on secondary texts of historians and social scientists who have since studied their works.

Documenting Brahmanism in the Colonial State: A Sociology of Resistance

The writings of anti-caste scholars and the leaders of non-Brahman movements described the entrenchment of Brahmanism in state institutions. These scholars conceptualized the state as both an independent organizational actor and one deeply embedded in society. Many who led or participated in these movements produced scholarship, created newspapers and magazines, and wrote manifestos that offered powerful critiques of caste.[57] They established an alternative corpus that criticized Brahmanical domination and British colonialism, and challenged the Brahmanical sociology that Nicholas Dirks names. Their scholarship—a sociology of resistance—highlights how caste elites and colonial administrators controlled knowledge-making institutions, and documents the formation of educational and occupational pathways previously inaccessible to caste-oppressed groups. Jotirao and Savitribai Phule, Iyothee Thass, B. R. Ambedkar, and Periyar, among others, critiqued how (*a*) the labor of caste-oppressed groups fueled the expansion of the colonial state, yet Brahmanism led to the exclusion of these workers from the institutions they built; (*b*) caste elites controlled the administrative bureaucracy of the colonial state; (*c*) Brahmanical ways of thinking and attitudes perpetuated inequitable policies and practices that dominated bureaucratic operations; (*d*) existing mechanisms of redress further perpetuated inequities; and (*e*) *swaraj* (self-rule) was impossible without the annihilation of caste. Each of these factors contributed to the institutionalization of Brahmanism in the state, which I detail below.

Oppressor Castes Control Knowledge-making Institutions

Caste elites dominated knowledge-making institutions. Writing in the 1870s in western India, Jotirao Phule describes how "the higher European officers generally view men and things through Brahmin spectacles," which prevented them from directly challenging Brahmanism.[58] These *Brahmanical spectacles* shaped how the colonial state approached its subjects and its power to name groups, their relative status to one another, the development and circulation of knowledge about caste, and the broader production of symbolic power.[59] At a time when the nationalists attacked British imperialism, Phule's writings also challenged Brahmanical teachings and historiographies of "Hindustan" that centered the arrival of "Arya Bhat-Brahmans" in the creation of India.

Phule provides an alternative golden age of India centering King Bali and Balisthan that contested elite constructions of Hindustan and an Aryan golden age.[60] Writing around the same period in south India, Iyothee Thass critiqued the Aryan theory of India's origins, arguing that Brahmans as Aryans entered the country and enslaved the non-Aryans or Dravidians, who lived in a free, prosperous, and egalitarian society before the Aryan conquest.[61] Both put forward an alternative story that decentered the role of Aryans and caste elites in the sociopolitical origins of the subcontinent.

Phule also documented the psychological power of Brahmanism perpetuated through mainstream books, publications, and media that helped caste elites maintain control of knowledge-making institutions. Phule argues that while "the shudras and atishudras have been freed from the physical slavery of the brahmans since the advent of the British … they still remain ignorant and captive in the mental slavery which the brahmans have perpetuated through their books."[62] Schools and books were key sites for reproducing (and potentially challenging) caste power. Formally exclusionary policies blocked oppressed-caste groups from accessing an education. Much of the organizing by early anti-caste activists was to create educational access for caste-oppressed groups.[63] Phule and Ambedkar, and other non-Brahmans, emphasized the democratizing potential of education—to develop students' critical consciousnesses to organize for change in the classroom and beyond.[64] Shailaja Paik discusses how Phule saw the crucial need for education to develop a "third eye," or a consciousness to challenge gendered caste oppression. He titled his first political play *Trutiya Ratna* (Third Jewel) (1855) as a metaphor for the third eye. Muktabai Salve, a student of Savitribai Phule, similarly wrote about the exclusion of oppressed castes—like her own Mang caste—from textual knowledge and pleaded with members of Mang–Mahar Dalit communities to open their third eye through education to challenge Brahmanism and patriarchy.[65]

Caste-oppressed groups advocated for access to schools free from Brahmanical ideologies.[66] Phule emphasizes this point when he says, "Let there be schools for the Sudras in every village; but away with all the Brahmin schoolmasters!"[67] Brahmanical ways of thinking exerted psychological control over Shudras and kept them in mental slavery. This powerful form of exclusion, along with formally exclusionary policies, helped to maintain oppressor-caste control of knowledge-making institutions, from schools and curriculum to the state census and mainstream media.

As discussed earlier, caste elites played a disproportionate role in analyzing and shaping the publication of caste-wise data from census. In the 1940s, Ambedkar argues, "[T]he Census of India has over a number of decades ceased to be an operation in demography. It has become a political affair."[68] Ambedkar was specifically referring to the decreasing proportion of SCs registered across two successive censuses and his disbelief that the relative size of the SC community was declining. He named the census a political project that underestimated the size of the SC population and in doing so limited the size of the reservation quota for SCs. Politics have always been deeply embedded in the production of census data, and Ambedkar recognized the social process of producing census data. Despite this reality and complexity, Ambedkar argued that caste-wise data should be collected in the censuses of independent India to understand caste hierarchy and to develop policies to annihilate it.[69]

Oppressed Castes Build State Institutions, yet Blocked from Accessing them

The labor of oppressed castes created the immense wealth of the British Empire and funded the development of state institutions and public works projects, which both foreign colonizers and caste elites enjoyed. In 1873, Phule writes in his book *Slavery*:

> It is an admitted fact that the greater portion of the revenues of the Indian Empire are derived from the Ryot's [oppressed castes"] labour—from the sweat of [their] brow. The higher and richer classes contribute little or nothing to the state's exchequer. A well-informed English writer states that, "Our income is derived, not from surplus profits, but from capital; not from luxuries but from the poorest necessaries. It is the product of sin and tears." That Government should expend profusely a large portion of revenue thus raised, on the education of the higher classes, for it is these only who take advantage of it, is anything but just or equitable.[70]

The labor of oppressed castes directly built and financed colonial institutions—from schools and public infrastructure to the courts. The caste-based forced labor regime in rural estates continued even after the formal abolition of slavery; the labor of Dalits generated immense wealth for landowners and fueled the expansion of the state.[71] At the same time, the institutions created by oppressed-caste labor remained inaccessible to them as colonizers and caste elites controlled access and related resources.

Even in large cities, caste-oppressed groups struggled to secure basic resources like water. Phule describes how, during a water shortage, government officials failed to provide water to caste-oppressed communities:

> Do you know, one of these "patriotic" *bhats* [priests/Brahmans] was the Chief Officer of the municipality last year when the position of water supply was very serious; yet he could not make arrangements for the atishudras [Dalits] to draw drinking water from the government tank! That is why there is a great need for at least one of the atishudras to be a member of the Municipality.[72]

Government officials, who were mostly Brahman, refused to provide water to Dalits. In a social system where Ambedkar describes "there is an utter lack" of "consciousness of kind," public officials (who are overwhelmingly caste elites) could not see the humanity of people deemed "untouchables"—blocking access to the water and other basic infrastructure that oppressed-caste labor built.[73]

The disconnect between oppressed castes' labor building public institutions and at the same time oppressed castes being excluded from their use occurred across contexts. In her research on education and Dalit girls in Maharashtra, Shailaja Paik argues that "with their toil and taxes, Dalits contributed to establishing and maintaining the systematic apparatus of education yet were virtually excluded from its benefits."[74] Describing the socio-educational reforms advocated by non-Brahman and Dalit radicals in early twentieth-century Maharashtra, Paik states that they conceptualized "Dalit exclusion from common schools as a barrier to both individual freedom and advancement, and to their collective ability to secure equal *maanavi hakka* (human rights) and *naagarikatva* (citizenship)."[75] Exclusion from the same institutions that their labor built negated the humanity and citizenship of oppressed castes.

Oppressor Castes Dominate the Bureaucracy of the Colonial State

During the colonial period, Brahmans dominated the government service with psychological and material consequences. In the early 1890s, the civil administration in India was "carried on by native agency, supervised by a small body of Englishmen"—with only about 750 British officials in the covenanted civil services, excluding the police.[76] Out of the 2,600 officials in the higher judicial and executive services, approximately thirty-five were Englishmen not domiciled in India, while 80 percent were Hindu—of which half were Brahman—and most of the remaining officials were Muslim.[77] As such, Brahmans were severely overrepresented (that is, approximately eight times their proportion of the population), while the representation of Muslims was slightly less than their proportion of the population.[78] In the lower, or subordinate civil services (of approximately 110,000 positions), 97 percent were Indian.[79] In 1918, in the princely state of Mysore (present-day Karnataka), Brahmans held 65 percent of gazetted posts and 70 percent of non-gazetted posts in the Mysore state services while forming 3.4 percent of the population according to the 1911 Census.[80] Brahmans held most positions in the upper ranks of government. Phule's writings from the 1870s to 1880s are filled with examples of how Brahmans and Brahmanical ways of thinking pervade the administrative bureaucracy and exert control over Dalits and Shudras in local government offices. He argues, "[O]ur government departments are infested with *bhat* [priests/Brahman] employees and this has had a terribly adverse impact on the *shudras* and *antishudras* [oppressed castes]."[81]

Movements against Brahmanism in the south and west pushed to diversify the administrative bureaucracy. Non- and anti-Brahman movements and Dalit movements in the early twentieth century built upon the contributions of Phule, Thass, Masilamani, and other nineteenth-century activists to challenge Brahmanical control of the administrative bureaucracy by creating access to local primary and secondary schools, developing targeted scholarships for higher education, and expanding the entry of oppressed castes into a wide range of government jobs. They argued that the colonial system of education, with its focus on elite higher education, neglected primary education for the masses, thus benefiting caste elites while continuing to prevent historically excluded groups from accessing education and government jobs. These movements led to the "reservation" of government jobs to diversify the composition of the administrative bureaucracy in the 1920s and 1930s in the south and west—first within princely states and regional governments and later in the central government. In princely Mysore, the Miller Commission classified all non-Brahmans as "backward" classes. Beginning in 1921, the "backward" classes had a 75 percent reservation in government jobs in Mysore, and this continued until 1959.[82] While the composition of the bureaucracy began to diversify during the colonial period, the ongoing dominance of Brahmans and caste elites had direct consequences on the development and continuation of unjust policies and practices.

Discriminatory Policies and Practices Pervasive in State Institutions

Brahmanical ways of thinking supported attitudes of disrespect, inferiority, hostility, and indifference toward oppressed castes and led to unjust laws and rules, as well as discriminatory norms and environments. Similarly, Brahmanism privileged caste elites who monopolized access to education, government jobs, and public and private resources. The administrative state perpetuated these "policies of inequity," as Ambedkar calls them, both formally and informally. Phule's and Ambedkar's writings emphasized how units of the state enabled untouchability, caste supremacy, and patriarchy through policies and practices. Caste elites restricted the entry of oppressed castes into the bureaucracy despite reservations, fostered attitudes of contempt and hostility toward oppressed castes in their daily work, and developed formal and informal policies to prevent the promotion of oppressed castes into the higher tiers of the bureaucracy.

Phule wrote extensively about the discriminatory practices that pervaded local government bureaucracies—in villages as well as larger cities like Pune during the colonial period.[83] He described how Shudras faced barriers and even lost access to land they previously toiled, in the process of securing documentation of land tenure, claiming property, paying taxes, or accessing loans in government offices. Even if they secured all the necessary documents, they still failed to achieve the outcome they desired. They were beholden to Brahman officers in all matters related to property and accessing other forms of collateral. Oppressed castes also knew that they had to pay extra to acquire services. Phule explains, "[T]hey know from their own everyday experience that the brahmans in all government offices really won't budge unless their palms are greased. So now they say, 'Never stir out of your houses without a lot of grease if you want to get your work done in government offices.'"[84] In the process of trying to access public services, oppressed castes experienced harm and violence, including the additional burden of paying bribes and facing daily indignities. Phule argued that the diversification of state offices was necessary to dismantle practices of exclusion.

Hostility toward oppressed castes permeated state organizational practices, social interactions, and social norms. In the preface to *Slavery*, Phule states how "if a Sudra or Atisudra repair to his Court, the treatment which he receives is akin to what the meanest reptile gets. Instead of his case receiving a patient and careful hearing, a choice lot of abuse is showered on his devoted head...."[85] Phule's analogy highlighted the antagonism and abuse that oppressed castes experienced in the courts and other state institutions. The discriminatory practices and policies perpetuated structural inequalities, physical harm, indignities, and psychological violence. The monopoly of state offices and public resources by caste elites—whether jobs, access to high-quality education, availability of clean drinking water, or land—created a discriminatory situation where public institutions and resources disproportionately benefited caste elites. Oppressed castes even "chose" to opt out of trying to access government services due to the hostile and discriminatory environment.

Along with open hostility, indifference also existed. Ambedkar described the disinterest in the legislature to challenging caste-based oppression as well as overt hostility.[86] Indifference was an active position and fully in line with the status quo—although in less obvious ways. The indifference of a legislature toward the concerns of "untouchables"—even when these legislators had been voted in to represent their constituencies—meant that change was unlikely in the absence of additional organizing by caste-oppressed groups. In not working to disrupt the ideologies and practices that perpetuated discrimination and indignities, political leaders allowed Brahmanism to continue unchallenged. Given the perniciousness of caste and the embeddedness of Brahmanism across institutions, Ambedkar argued that social reform (that is, destroying caste hierarchy) was a necessary precondition for political equality.

Mechanisms of Redress Further Perpetuate Inequities

Oppressed castes faced additional burdens when seeking redress for discriminatory treatment. In *Slavery,* Phule discusses at length why Shudras did not report the misdeeds of oppressor castes to government officials.[87] Phule begins by describing the structural challenges to seeking a remedy: "You expect the ignorant plaintiff who doesn't even know how to write the letter 'a' to stand before the great officer and present to him his grievances in a systematic way?"[88] The intergenerational discrimination and exclusion that oppressed castes experienced—including the lack of access to schooling and literacy—made it difficult to challenge the actions of government officers or issue a complaint before the local arbitrator. Phule explains that even when oppressed castes made a formal complaint, and a judge listened, the entire bureaucracy then mobilized to create confusion such that the judge would "scold the poor complainant for being a fuzzy nitwit and send him packing home with a few well-chosen epithets!"[89]

The bureaucratic system blocked the possibility of a fair outcome for those who sought redress. Most likely, the process was filled with further injustices that included tormenting and reframing the actions of the person who filed the complaint as the cause of the problem. At best, administrators ignored the complaint and the status quo continued, as if the person had never filed the complaint. Their effort, however, depleted limited time and resources. Phule extensively documented how institutional mechanisms that existed to correct injustices or offer arbitration perpetuated indignities. The grievance redressal system protected the interests of oppressor castes and caused further violence and harm to oppressed castes.[90] More than seventy years after Phule's account, Ambedkar similarly describes how state institutions continued to pursue their "policy of inequity towards the Untouchables without any curb" and how "untouchables" have "no way of escape from the destiny of degradation."[91] Efforts to dismantle Brahmanism within the administrative state and related institutions by filing individual claims through formal mechanisms were largely unsuccessful—fostering a situation where few utilized mechanisms to register complaints or seek redress. External organizing remained crucial for transformative change.

Caste Hierarchy Remains Foundational to Swaraj

Throughout the colonial period, anti-caste leaders argued against the possibility of self-government without dismantling internal caste hierarchy. Iyothee Thass critiqued Indian political leaders, asking, "[W]ill they make pariahs governors and the brahmins army commanders? Never. Instead, they will argue that the low castes cannot govern, while the high caste brahmanas are not trained to fight."[92] Thass denounced the *swadeshi* movement and argued that the spirit of the movement was based on caste pride, religious pride, pride of knowledge, and pride from wealth.[93]

These critiques highlighted shortcomings in the Congress's elite composition, policies, and internal ideologies. As the party expanded its scope and reach in the early twentieth century, it failed to disrupt the agrarian order, popularize land reform for peasants, or support anti-landlord struggles. Gail Omvedt's analysis of Jawaharlal Nehru's writings finds that the "father of secularism" feared that destroying caste would wreak havoc on the social order which centered the group unlike western individualism.[94] Gandhi's agenda focused on driving out western civilization and its materialism using non-violent militancy; cleansing the mind and body through fast, meditation, and physical labor; controlling one's mind and will; and having Dalits gain self-respect without empowerment. Gandhian nationalism envisioned a decentralized village India where social and cultural life thrived and economic production was sufficient for a life free of materialism.[95] Ambedkar's push for a separate electorate for "untouchables" in the 1920s and 1930s clashed with Gandhi, who advocated for a gradual reform to the caste system. In response, Gandhi threatened to starve until death and started a fast until a compromise was reached.[96] Congress leaders sided with Gandhi and believed that the salience of caste and other traditional identities would steadily decline over time as a unifying Indian identity strengthened. As such, they failed to support radical reforms to destroy caste hierarchy.

More than seventy years after Phule and Thass critiqued oppressor-caste control of the bureaucracy, Ambedkar continued to fear the power of caste elites as he drafted a legal framework for independent India.[97] Without dismantling caste hierarchy, Ambedkar foresaw that "self-rule" would create a state "administration unbridled in venom and in harshness, uncontrolled by the Legislature and the Executive, [able to] pursue its policy of inequity towards the Untouchables without any curb."[98] For Ambedkar, *swaraj* was not a moment of liberation but one that would lead to the expanded tyranny of caste elites over "untouchables" as the full apparatus of the state came under the control of oppressor castes. Ambedkar argues:

> ... what can *Swaraj* mean to the Untouchables? It can only mean one thing, namely, that while today it is only the administration that is in the hands of the Hindus, under *Swaraj* the Legislature and Executive will also be in the hands of the Hindus. It goes without saying that such a *Swaraj* would aggravate the sufferings of the Untouchables.[99]

The exit of British officers would expand the dominance of caste elites, and Ambedkar feared that oppressor castes would take over the state apparatus and further institutionalize the caste-based discrimination already pervasive in the bureaucracy. He states, "[U]nder Swaraj the Untouchables will have no way of escape from the destiny of degradation which Hindus and Hinduism have fixed for them."[100] While nationalist leaders in India celebrated ideals of justice, equality, and fraternity—which are antithetical to caste—institutional change remained limited in decolonizing India. Thus, Ambedkar continued his work of developing and expanding policies and programs to counter Brahmanism.

*

The writings of anti-caste leaders during the colonial period documented the embeddedness of Brahmanism in the state and foretold why efforts to annihilate caste would be a low priority in independent India. Caste elites would have a stronghold in the administrative state at a time when political power based on caste hierarchy increasingly contradicted the egalitarian philosophy of a democratic society. The census became one of several state projects where the battle over the visibility of caste hierarchy and power took place.

Caste and Census in Independent India

> The forthcoming census is the first census of a Free Republican India. Formerly, there used to be elaborate caste tables which were required in India partly to satisfy the theory that it was a caste ridden country and partly to meet the needs of administrative measures dependent of caste divisions. In the forthcoming census, this will no longer be a prominent feature and we can devote our energies and attention to the collection and formulation of basic economic data ... of the individual and the state.
>
> —Sardar Vallabhbhai Patel, 1950[101]

In the lead-up to the 1951 Census, Congress Party leaders removed a caste-wise enumeration from India's first post-independence census. The decision to stop the colonial practice of enumerating caste was ideological. Sardar Vallabhbhai Patel, deputy prime minister of India, and other Congress leaders envisioned a more restricted role for caste in state administration, arguing that previous caste counts had reinforced caste boundaries and, in doing so, facilitated colonial rule through an emphasis on internal divisions. Prime minister Nehru stated at a Tribal Conference in Delhi in 1952 that "the greatest problem of India today is psychological integration and consolidation, to build up a unity which will do away with provincialism, communalism and various other 'isms' which disrupt and separate."[102] Congress leaders wished to develop a national identity against the backdrop of multiple forms of internal diversity, including religion, language, caste, and class. In the 1951 Census, then, caste became defined more narrowly as solely a source of disadvantage, while privilege did not require documentation. By restricting data collection to caste-oppressed groups, decennial censuses helped to "unmark" caste power and obscure inherited caste privilege in independent India.[103] Census policy on caste propagated an ideology of castelessness, which narrowed caste to certain aspects of disadvantage, ignored the lived experiences of unenumerated caste-oppressed groups (such as Dalit Muslims and Christians), and masked caste privilege.

Yet, while Congress leaders sought to promote a seemingly universal Indian identity and become credible all-India representatives, they took a conservative approach to challenging Brahmanism and caste hierarchy—failing to pass policies that would annihilate caste. Instead, caste-based advantages and privilege became reconfigured as "merit" based on individual effort and capabilities, or a good family background—a common euphemism for caste privilege.[104] There was minimal attention at the policy level to how caste privilege continued to translate into structural advantage. The Congress leadership perpetuated castelessness in the census and related institutions following independence in 1947. Political scientist Steven Wilkinson traces how a series of decisions in 1949 by the Constituent Assembly charged with drafting the Constitution of independent India brought to life Nehru's goal "to make the Indian state 'color blind.'"[105] The Constituent Assembly dismantled reservations in legislatures and government employment for Muslims, Christians, and Sikhs and eliminated separate electorates for Muslims. Nehru also wanted the Assembly to abolish reservations for SCs, STs, and Anglo-Indians, but Ambedkar successfully advocated to extend reservations in his role as chair of the drafting committee of the Indian Constitution and later as law minister. While Congress claimed minorities requested the abolition of reserved seats and separate electorates due to their shared allegiance to a secular India, scholars and journalists have documented how Congress leaders warned religious minorities to remain silent and convinced a recent defector from the Muslim League to speak in support of a general electorate—and she did so in a reluctant manner.[106] Representatives of the Muslim League later contradicted Congress's assertion that minority leaders had assented to the removal of reserved seats and separate electorates.[107] The Constituent Assembly also backpedaled on Congress's promise to protect Urdu speakers. Muslim League member Z. H. Lari resigned from the Assembly "to express resentment at the decision taken by that body to replace Hindustani by Hindi as the 'lingua franca' of India and to adopt the Devanagari script to the entire exclusion of the Urdu script."[108] Several Muslim legislators quit to protest a process that was dismantling minority rights. The government also envisioned and implemented a limited role for caste in the census—first announcing in December 1949 that it would no longer produce caste-wise data.[109] Wilkerson argues that "this last step was not just a symbolic rejection of caste; it removed the data that underpinned ethnic quotas in pre-Independence India."[110] Following the passage of the 1949 Census Act, the government created a permanent Office for the Registrar General & Census Commissioner, India (ORGI) within the powerful Home Ministry.[111] The ORGI and the Home Ministry now had the authority to decide what topics and questions to include (or exclude) in the decennial census. The ORGI scaled back the collection of caste data in the 1951 Census to "special groups," consisting of SCs, STs, and Anglo-Indians—the three groups that still had reservation benefits in 1951. The stated justification to continue data collection on religion was to enumerate SCs—as only Dalits who identified as Hindu could be enumerated as SC in 1951. All censuses in independent India have enumerated SCs and STs, for whom parliament has extended reservations every ten years.[112]

The definition of the SC category excluded non-Hindus in the 1951 Census. Other Dalits—including Buddhists, Sikhs, Muslims, Christians, individuals who followed a range of spiritual practices outside of the Hindu fold, agnostics, and atheists—were not enumerated as SC. The decision to limit the category of SCs to Hindus involved a complex set of negotiations that aligned with the broader project of nationalism among Congress leaders that shifted attention away from caste hierarchy and supported the making of a "Hindu majority." In the all-India census report of 1951, there was virtually no mention of caste nor caste tables, and the report focused on economic and employment data. This stood in stark contrast to the 1931 Census, or the last census that had published caste wise data, where the all-India report and regional reports included extensive tables on caste. In the 1951 Census, the data collected on SCs and STs (and in some states Anglo-Indians) were published in a single table in provincial or state-level reports.[113] The restricted enumeration of caste also supported the creation of a "Hindu majority" with the limiting of SCs to Hindus. This decision ignored both the complex landscape of religious traditions of caste-oppressed groups and the historical aversion of oppressor castes to conceptualize so-called untouchables as part of the Hindu community.[114] In his study of the Lal Begi tradition, anthropologist Joel Lee argues against popular and scholarly assumptions that Dalits "insofar as they are not avowed converts to Buddhism, Islam, or Christianity, somehow *belong* to the Hindu community and should therefore see themselves as Hindus."[115] The construction of the "Indian" was strongly associated with the formation of a Hindu cultural identity—one which sought to superficially include SCs and STs in the creation of a political majority—but with caste elites as the public face and in control of institutional power.

While the SC category later expanded to include Sikhs in 1956 and finally Buddhists in 1990, Dalit Muslims and Dalit Christians remain ineligible. Sikh leaders had negotiated for the inclusion of Dalit Sikhs in the SC category in 1949 when Congress leaders pressured religious minorities to give up the demand for separate electorates and reservations.[116] Muslim and Christian leaders did not negotiate for the inclusion of Dalit Muslims and Dalit Christians within the SC category during the 1949 debate.[117] The Constituent Assembly's removal of reservations for religious minorities in legislative assemblies and government employment occurred alongside the political leadership's refusal to support "backward"-caste leaders trying to expand reservations in government jobs in the 1950s. The home minster rejected the request, arguing that caste was "the greatest hindrance in the way of our progress toward an egalitarian society, and the recognition of specified castes as backward may serve to maintain and perpetuate the existing distinctions on the basis of caste."[118]

In the 1961 Census and continuing to the present day, the ORGI has enumerated SCs and STs, while categorizing everyone else as "other" or "non-SC/ST" by default (see Table 2.2). While the collection of caste-wise data in the census became limited to data required for the administration of affirmative action benefits in the early post-independence period, this explanation became inadequate following the extension of reservations to OBCs in the central government by the early 1990s. Commissions, caste-based organizations, and state governments have pushed for the inclusion of a full caste count in the census but have had little success due to the entrenchment of castelessness within the state and mainstream understandings of Indian nationalism propagated by the Congress and BJP.

Efforts to Restore a Caste-wise Enumeration in the Census

Calls to reintroduce the collection of caste-wise data emerged from different corners following the 1951 Census. The first Backward Classes Commission (BCC), chaired by Kaka Kalelkar, had a mandate to develop criteria to identify socially and educationally "backward" classes, create a list of groups based on the developed criteria, and recommend what the state should do "to remove such difficulties or to improve their conditions."[119] Appointed by the president of India, the commission gathered existing data within the administrative bureaucracy and also developed a detailed questionnaire for state governments to complete.[120] One specific area of inquiry on the questionnaire related to the inclusion of caste in the decennial census.[121] The commission asked state governments, "Do you think that the abolition of the mention of caste or sub-caste in the census of 1951 has been useful or otherwise for determining the condition of backward classes? What procedure would you recommend for adoption in the future censuses?" The commission tried to get caste data from the ORGI, as well as the Education Department and National Sample Survey, but found that available data were insufficient.[122] Due to the difficulty in finding administrative data that could help define "backward" classes, the BCC's first recommendation in its 1955 report was titled "Census and Caste" and called for a full caste-wise enumeration as part of Census 1961:

Table 2.2 Caste in the Census in Independent India (1951–2001)

Year	Census Questions	Notes
1951	Q2: Nationality, Religion and Special Groups.– Part (a). Nationality– Write I for all Indian Nationals. For others, write Nationality in full. Part (b). Religion– Write H for Hindu; M for Muslim; C for Christian; S for Sikh; J for Jain. For others, write the answer as actually returned. Part (c). Special Groups	• The question on religion had slight variations across regions. For example, in Bombay the question added the option of Parsi: "write 1 for Hindu, 2 for Muslim, 3 for Christian, 4 for Sikh, 5 for Jain and 6 for Parsi." In addition, while previous censuses had included tribal religions for this question, the rewording of this question dramatically reduced this classification. For example, in Bombay state no one was recorded as having a tribal religion while in the 1931 census 200,586 people were recorded as having a tribal religion.[1] • The question related to caste and tribe (that is, Part c. Special Groups) had tailored regional instructions for enumerators. Enumerator instructions specified not to record the caste affiliations of persons who did not belong to Scheduled Castes, Scheduled Tribes, or Backward Classes (sometimes described as "non-backward castes" or "non-scheduled castes"). One example from Punjab:[2] ○ Enumerator instructions: "Part (c). Special groups–A person is free to return whatever caste he likes or to say that he is not a member of any caste or tribe. Write 'A' for Anglo-Indian '0' for any person who says that he is not a member of any caste or tribe and '1' for any person who is a member of anyone of the castes mentioned in the list of non-backward castes supplied to you. In all other cases, write: the name of the caste or tribe as returned by the person enumerated." ○ Instructions for training enumerators: "A list of non-backward classes has been prepared for the Punjab zone. For a person who says he is not a member of any caste, numeral '0' be noted down in Part (c) of Question No. 2. For a person who belongs to one of the castes mentioned in the above-mentioned list of non-backward classes numeral 1 as abbreviation be noted. For an Anglo-Indian use abbreviation 'A'. In all other cases the name of the caste or tribe as returned by the person enumerated be noted down. The following classes in the Punjab zone are known as Harijans or those suffering social disabilities or such as are regarded as backward:" [list of 63 scheduled castes provided].

(*Continued*)

Table 2.2 (*Continued*)

Year	Census Questions	Notes
1961	Q5 (a) Nationality 5 (b) Religion 5 (c) S.C./S.T.	• The enumerator instructions for the nationality (Q5a) and religion (Q5b) questions remained the same as in the previous census (1951), with the addition of 'Budhist' to the list of religions specified. • The question related to caste (Q5c) narrowed to enumerate Scheduled Castes and Scheduled Tribes (and not a broader category of special groups like in Census 1951). The enumerator instructions were detailed: ◦ "The answer to this question will be recorded only if a person belongs to a Scheduled Caste or Scheduled Tribe. If the person belongs to a Scheduled Caste or Scheduled Tribe obtaining in your state, or in your district write the name of the Caste or tribe to which he belongs. For all others write 'x' in this column. A list of Scheduled Castes and Scheduled Tribes obtaining in your State or in your district is given elsewhere in these instructions. ◦ If the person belonging to a Scheduled Caste or Scheduled Tribe returns his caste or tribe by a synonym or a generic name of a caste or a tribe as given in your list, write the name as returned and the name of the relevant caste or tribe, as per printed list, within brackets. ◦ Do not write the names of Scheduled Castes in general terms as *Harijan*, *Achhut*. You should ascertain the name of the caste when it is returned and write it. If a person is negligent and insists on calling himself merely *Harijan*, tell him that this description will not earn the person any benefits under the Constitution permissible to Scheduled Castes and Scheduled Tribes. This may persuade him to give out the correct name. ◦ Scheduled Caste can belong only to the Hindu or Sikh religions. If a person belongs to a Scheduled Caste, there will be either 'H' or 'S' in the answer to question 5 (b). Scheduled Tribes may belong to any religion."[3]
1971	Q10: Religion Q11: S.C. or S.T.	• The religion question (Q10) remained similar to Census 1961. The enumerator instructions additionally state that "if a person says that he has no religion it may be recorded accordingly. Do not mistake religion for caste which will not be recorded here."

(*Continued*)

Table 2.2 (*Continued*)

Year	Census Questions	Notes
		• The caste question (Q11) remained identical to Census 1961.
1981	Q8: Religion Q9: Whether S.C. (1) or S.T. (2) Q10: Name of caste/ tribe	• The religion question (Q8) remained similar to the last two censuses. The enumerator instructions additionally state that "You should also not try to establish any relationship between religion and mother tongue." • The caste questions (Q9 &10) remained identical in intent to the last two censuses; however, the question now split into two parts, with the first question establishing whether the person is SC, ST, or "non-SC/ST" by default, and the second question recording the specific caste or tribe for SCs and STs (which must match a caste/tribe listed on the state-level SC/ST list).
1991	Q8: Religion Q9: Whether S.C. (1) or S.T. (2) Q10: Name of Scheduled Caste/ Tribe	• The religion question (Q8) remained similar to the last three censuses. • The caste questions (Q9 &10) remained identical to Census 1981.
2001	Q7: Religion (write name of the religion in full) Q8: If Scheduled Caste, write name of the Scheduled Caste from the list supplied. Q9: If Scheduled Tribe, write name of the Scheduled Tribe from the list supplied.	• The religion question (Q8) remained the same as the past four censuses. The enumerator instructions additionally emphasize that religion is an individual identity and that people in the same household can have different religious identities: "it is not necessary that all the members in the household profess the same religion. Therefore, enquiry should be made for each member of the household independently. We should not presume that the religion of the head or the respondent is necessarily the religion of every member being enumerated in the household." The instructions also mention that "you should not mistake religion for caste names." • Scheduled Castes now also include individuals who are Buddhist (along with Hindus and Sikhs as in past censuses).

Source: Natarajan, *Indian Census through a Hundred Years*, 630–81.

Notes:

1 J. B. Bowman, *Census of India, 1951, Vol. IV Bombay, Saurashtra and Kutch: Administration Report* (Bombay: Government Central Press, 1953), 13.

2 Lakshmi Chandra Vashishta, *Census of India, 1951, Vol. III Punjab Sub Zone: Administrative Report* (Simla: Controller of Printing and Stationery, 1951) 59, 65.

3 Natarajan, *Indian Census through a Hundred Years*, 655–66.

> Before the disease of caste is destroyed all facts about it have to be noted and classified in a scientific manner as in a clinical record. To this end we suggest that the 1961 Census be remodeled and re-organized so as to secure the required information on the following lines: (1) The Census operations should be conducted as a well-equipped continuous organization competent to supply information on various topics of sociological importance. (2) The Census officers must have permanent ethnologists and sociologists, in addition to the economists attached to them. (3) As long as social welfare and relief have to be administered through castes, classes or groups, full information about these groups should be obtained and tabulated.[123]

The BCC asserted an ideological difference with the political leadership in this recommendation; if the state was to play an active role in destroying caste hierarchy, full information on caste was required at regular intervals. The BCC identified 2,399 "backward" classes, which accounted for approximately 32 percent of the population.[124]

The Congress-led central government ignored the BCC's recommendation and did not include a full caste count in the 1961 Census. The political leadership argued that the commission had not applied objective tests for its identification of "backwardness" and therefore the identification of "backward" classes was suspect, as were its recommendations of how to improve the conditions of these communities. At the same time, the wide-reaching nature of the report, including measures that would have transformed rural property holdings throughout India, was not met favorably by Nehru and his cabinet, who had crafted development programs based on gradual and limited income redistribution, instead of an overhaul of rural relations.[125] In addition, the political leadership argued that the large number of communities that the BCC found "backward" undermined its usefulness.[126]

Supporting the Congress leadership's wide dismissal of the 1955 report was a lengthy cover letter submitted by chairman Kalelkar to prime minister Nehru, in which Kalelkar repudiated the contents of the report that he had signed. Kalelkar, a Gandhian, dissented from the report for fear that caste-based quotas would "destroy the unity of the nation and narrow down the aspiration of the people."[127] The home minister also offered a strong critique of the report when it was put before parliament in September 1956.[128] The only Kalelkar Commission recommendation that the central government implemented was to support the expansion of efforts to serve "backward" classes at the state level, which already existed in several southern states and a few other regions.[129] The central government delegated the task of identifying and improving the conditions of OBCs to state governments.[130] However, despite this delegation, they tried to steer state governments away from caste-wise data and toward economic measures. In the home minister's letter to state governments, he wrote that "while the State Governments have the discretion to choose their own criteria for defining backwardness, in the view of the Government of India it would be better to apply economic tests rather than to go by caste."[131] Over the next several decades, state governments began to expand their organizational apparatus for identifying and serving the needs of "backward" communities.

Organizations created to serve "backward" classes began or continued their work. In Mysore, the first BCC headed by R. Nagana Gowda submitted its report in 1960.[132] It prepared lists for groups deemed socially "backward," educationally "backward," and lacking adequate representation in state jobs. The state-level commission estimated that these groups accounted for 57 percent of the population. The commission also strenuously complained that it lacked sufficient data on literacy and other socioeconomic indicators and urged the state government to pressure the center to collect detailed caste-wise data in the 1961 Census. This started a consistent and early trend from commissions, state governments, courts, and caste associations requesting more extensive caste-wise data from the ORGI for effective policymaking.

In the absence of centralized data, state governments conducted ad hoc surveys and, in a few cases, detailed state-wide surveys to help create and maintain their OBC lists. Some of the practices and politics of the colonial period re-emerged in these efforts. For example, those in formal political power mediated the decision of which groups should belong on Karnataka's list of OBCs. Vokkaligas and Lingayats, two broad social categories that consist of numerous *jatis*, have controlled formal politics in Karnataka since it was created in 1956. These groups along with other non-Brahman castes had access to reservations from 1921 to 1959, and benefited from decades of non-Brahman reservations. As such, the inclusion of these two groups on Karnataka's backward classes list in the aftermath of the 1984 socioeconomic caste survey reflected their power within state government and the continuation of a stream of thought that equated caste power with Brahmans versus Brahmanism.[133]

Between the 1960s and 1980s, demands for the reintroduction of a full caste count in the decennial census sprang up from regional political parties and social movements hoping to defend or extend state-level reservations, and often in relation to legal cases concerning affirmative action programs.

Most of these early demands for a full caste count came from activists and politicians in the south and west, where reservation benefits were initially implemented. In 1951, the regions of Bombay, Madras, Mysore, and Travancore-Cochin had extensive programs for OBCs. Other regions had educational fee concessions, post-matriculation scholarships, and other types of educational programs for OBCs (for example, in Assam, Bihar, Uttar Pradesh, West Bengal, Rajasthan, Ajmer).[134] The definition of OBCs and their proportion of the population varied significantly by state.[135] By the mid-to-late 1970s most states had some type of reservations for backward classes in government posts.[136] However, states differed considerably on the scope of reservations in higher education, the size of the reservation quota in government jobs, and whether household income limited access to OBC reservations. For example, in the late 1970s, Karnataka, Kerala, Tamil Nadu, and Andhra Pradesh had 30–40 percent reservations in government positions, while other states such as Gujarat, Punjab, Himachal Pradesh, and Maharashtra had 4–15 percent reservations in government jobs.[137] The decentralized demands for a caste census that occurred between the 1950s and 1970s paralleled the political mobilization of OBCs within particular regions of the country and the expansion of state-level agencies and commissions to address the needs of OBCs. Many of the demands for detailed data on caste came from political parties or caste associations representing specific groups who believed that they were not getting their fair share of the reservation quota. Individuals and organizations representing groups who were excluded from reservations or unable to access reservations despite being eligible filed several landmark cases, which put additional pressure on state agencies to develop objective and legally defensible criteria on how they determined "backwardness."[138] During this period, the political leadership of the Congress effectively deflected the issue of collecting caste-wise data to state governments.

Twenty-four years after the first BCC at the national level came out with its report, prime minister Morarji Desai of the Janata Party (which was the first non-Congress party in power at the center after prime minister Indira Gandhi's unpopular Emergency period) appointed a second national-level BCC in early 1979.[139] The second BCC also advocated for a full caste count in the census. The commission, chaired by B. P. Mandal, was charged with developing criteria to identify "Other Backward Classes" and applying these criteria to generate a nationwide list of OBCs. The second commission took special care to tap several independent sources of data while creating its criteria, in light of the long-standing critique of the first commission's findings.

The Mandal Commission, as it became popularly known, designed indicators to measure backwardness and, in the process of doing so, tried to acquire caste-wise data. Mandal writes in his introduction to the commission's final report:

> We had to face enormous difficulties in the absence of caste enumerations figures after the 1931 Census. To avoid such difficulties in the future, a reference was made by me in my letter of 15th June, 1979 and 18th August, 1979, addressed to S/hri H.M. Patel and Y.B. Chavan, respectively. I had also requested Giani Zail Singh, Home Minister, to this effect in my letter of 31st March, 1980. I was informed that it had been decided that caste enumeration will not be carried out during the 1981 Census and that the present policy of not having enumeration of caste in Indian Census, will be continued, which needs reconsideration.[140]

Mandal wrote to three different home minsters during a one-year period, and each consistently refused to include a caste-wise enumeration in the 1981 Census. The commission instead used projections based on 1931 census data to estimate the total size of "backward" communities and identified 52 percent of the Indian population as "backward." It also identified 25 percent of the population as "forward" (see Table 2.3). The Mandal Commission recommended a 27 percent reservation quota for OBCs—equal to half its population estimate. Articles 15(4) and 16(4) of the Indian Constitution legally obligated the government to keep the reservation quota under 50 percent. The recommended OBC quota of 27 percent along with the SC and ST combined quota of 22 percent fixed the reservation quota in the central government below the required 50 percent.[141] The commission also recommended that a full caste count take place in the decennial census—despite the home ministry's resistance to Mandal's multiple requests for data.[142]

The ORGI should have minimally extended data collection to OBCs in light of the Mandal Commission report. But Indira Gandhi had returned to power by the time of the publication of the report in 1981; over the subsequent decade while she and her son, Rajiv Gandhi, led the Congress Party, they ignored the commission's recommendations. Following in Nehru's tradition, his daughter and grandson distanced themselves from a full caste-wise enumeration and an expansion of central government reservations based on criteria that focused on caste. While reservations for OBCs flourished at the state level, affirmative action benefits in central government jobs, educational institutions, and legislative bodies remained restricted to SCs and STs.

It once again took a change in leadership at the center for OBC reservation policies to progress. Soon after prime minister V. P. Singh of the Janata Dal came

Table 2.3 Mandal Commission's Estimate of Indian Population by Caste (1980)

Group Name	Percentage of Population
Scheduled Castes (SCs) and Scheduled Tribes (STs) Total	22.56
SCs	15.05
STs	7.51
"Forward" Castes Total	25.34
Hindu Castes and Communities	17.58
Non-Hindu Castes and Communities	7.76
"Backward" Castes Total	52.10
Backward Hindu Castes and Communities	43.70
Backward non-Hindu Castes and Communities	8.40

Source: Backward Classes Commission (B. P. Mandal, Chair), *Report of the Backward Classes Commission*.

to power as part of the National Front coalition government from 1989 to 1990, he issued an order to expand reservations to OBCs.[143] The political leadership extended affirmative action benefits to OBCs through a 27 percent reservation in central government jobs, as recommended by the Mandal Commission. This policy change created quite a controversy, leading to riots in parts of the country. Along with public backlash from elites, several petitions were filed before the Supreme Court questioning the legality of the policy and to stay its operation. After a two-year stay, the Supreme Court of India gave a landmark judgment in support of implementing the order in November 1992 and directed the central government of India, state governments, and administrations of union territories to constitute a permanent body to entertain, examine, and make recommendations regarding inclusion, exclusion, and under-inclusion in the list of OBCs.[144] The Government of India also directed the specification of socioeconomic criteria to exclude comparatively privileged individuals or groups from the OBC category—or the so-called creamy layer. During the legal debate, those for and against the Mandal Commission's recommendations called for more data on caste. While critiquing the methodology of the Mandal Commission report, sociologist Avatthi Ramaiah argued:

> The government should make a national level survey immediately and list out all the caste groups including that of the forward castes, and their socio-economic and educational status. The survey should take notes of all the technical errors found in the various criteria adopted by Mandal and formulate more rational criteria and scientific approach toward identifying the really deserving people within the castes and communities of each stratum of our society.[145]

Despite the expansion of reservations to OBCs, the executive bureaucracy remained resistant to collecting caste-wise data. Yet calls for updated data on caste persisted, as courts, government commissions, political parties, and caste-based organizations continued to make demands for "objective" data. In several legal decisions in the aftermath of implementing reservations for OBCs, the Supreme Court stated that updated data on caste were needed if states wanted to further extend OBC reservations.[146]

Since Census 2001 was the first decennial census following the expansion of central government reservations to OBCs, it should have also collected data on OBCs for the implementation of reservations. Census commissioner M. Vijayanunni organized numerous meetings with data users and the public during his tenure between 1995 and 1999. These encounters with diverse constituencies convinced him that state and local governments and policymakers required caste-wise data to effectively implement existing laws and policies, including reservations and other efforts to redress structural inequalities.[147] Vijayanunni believed that only the census bureau had the technical expertise and resources to undertake the task. For his Ph.D. dissertation, he had studied censuses in British India and understood the complexity of the gargantuan task of collecting and processing caste data from every individual in the territory. Vijayanunni thought that available technology could significantly ease the burden of processing and aggregating caste-wise data.[148] Along with Vijayanunni, the Ministry of Social Justice also advocated to include an expanded caste count in the 2001 Census. The Ministry of Social Justice and Empowerment, National Commissions for Scheduled Castes and Scheduled Tribes, BCCs (at the state and national levels), and related units had advocated for a caste-wise enumeration in the census to support the development and administration of justice-oriented policies and programs.

Although a caste-wise enumeration in the census made sense for the administration of reservation benefits, Vijayanunni, a life-long civil servant in the IAS, or the elite cadre of government administrators, understood the nature of institutional resistance. Vijayanunni expected that the strongest opposition would come from within the ORGI itself. The top tiers of the central bureaucracy remained dominated by caste elites, as detailed in Chapter 1. Vijayanunni's term as census commissioner ended in 1999—just prior to the roll out of Census 2001. His successor maintained the status quo in census policy on caste. Relatedly, the home minister at the time, A. K. Advani, was staunchly opposed to an expanded enumeration of caste. Advani held one of the most powerful ministerial posts in the BJP-led National Democratic

Alliance (NDA) government.[149] The BJP did not want to implement a change to census policy on caste, and the BJP leadership blocked further debate on the subject within the government. Therefore, in the first census after the expansion of OBC reservations in the central government, the caste questions remained unchanged. Political resistance to the enumeration of caste in the census had succeeded once again.

Looking Back and Ahead

In colonial censuses, Brahmanical perspectives became the basis of the official census rankings of caste and heavily shaped the state's understandings of caste. Yet this period was also crucial for organizing against caste and Brahmanism, and census data (including caste-wise data) supported the demands of caste-oppressed groups for self-representation in legislatures and government service, expanded access to education and other public services, the end of untouchability, and the fostering of anti-caste subjectivities, knowledge-making spaces, and policies to dismantle Brahmanism. Contemporary scholars and activists who point to colonial censuses as evidence for why caste-wise data should not be collected in twenty-first-century censuses focus on scholarship that emphasizes the role of caste elites in the construction of colonial knowledge and the use of census data to inform colonial policies of "divide and rule." Equally important during this period is an extensive sociology of resistance that emerged from anti-caste leaders and movements that used caste-wise data (and its intersections with social, economic, and demographic data) to make claims on the state as political subjects with the right to self-representation, education, secure livelihoods, dignity, and equal respect. In addition, the colonial census through trial and error developed a methodology for enumerating caste. Regional census offices played a key part in processes of commensuration—by creating caste lists prior to the census and developing additional instructions for enumerators to orient them to the complexity of enumerating caste. Their preparatory work highlights the detailed planning and coordination required between regional governments and the center to enumerate caste. In addition, these historical censuses reveal the consideration and planning needed to enumerate caste across religious communities and the importance of all-India categories for the successful commensuration of caste in the decennial census.

Nehru, Gandhi, and other Congress leaders in the 1930s and 1940s minimized attention to caste, tried to do away with reservations, and sought to enumerate Dalits within the "Hindu majority"—in doing so they strengthened an ideology of castelessness in political life.[150] The Congress leadership removed a full caste-wise enumeration from the census after independence, while the limited enumeration of SCs and STs continued to determine the size of each group's reservation quota. An ideology of castelessness became embedded in the decennial census as caste became narrowed to a source of disadvantage while the systems of power and privilege that created this disadvantage were no longer enumerated or made visible.

Efforts to (re)introduce a caste-wise enumeration in the censuses of independent India have proved unsuccessful. The central government deflected the issue of a caste-wise enumeration in the census and the identification of broader groups of "backward" communities to state-level commissions and provincial governments. While political leaders critiqued the criteria used by two national-level BCCs to identify "backward" groups, they refused to implement the recommendations of the commissions to collect caste-wise data in the census for an improved process of identifying "socially and educationally backward" communities. Despite the eventual expansion of central government reservations to OBCs in the 1990s, the bureaucratic need for updated caste data remained insufficient fodder for an expanded caste count in Census 2001. The executive bureaucracy continued to block a caste-wise enumeration in the decennial census. Yet, as we will see in the next chapter, the organizing that took place in the lead-up to Census 2011 finally posed a challenge to the status quo.

Notes

1. M. K. Gandhi, "India Round Table Conference (Second Session): Proceedings of Federal Structure Committee and Minorities Committee," in *Collected Works of Mahatma Gandhi (13 October 1931–8 February 1932)*, vol. 54 (New Delhi: Publications Division Government of India, 1999), 159, http://www.gandhiashramsevagram.org/gandhi-literature/mahatma-gandhi-collected-works-volume-54.pdf, accessed April 1, 2024.
2. Vivek Kumar Singh, "Counting Caste: A Step toward Radical Anticasteism," *Prabuddha: Journal of Social Equality* 7, no. 1 (2022): 1–13.
3. Gandhi, "India Round Table Conference (Second Session)," 158.
4. Decolonizing is in quotes to acknowledge the mainstream nationalist movement's effort to decolonize from British colonial rule but limited commitment to challenge internal caste domination.
5. Castelessness frames caste as a problem of oppressed groups and conceals caste privilege. S. Deshpande, "Caste and Castelessness."

6. G. S. Ghurye, *Caste and Race in India* (London: K. Paul, Trench, Trubner & Company, Limited, 1932); M. N Srinivas, *Caste in Modern India and Other Essays* (London: Asia Publishing House, 1962). Cohn, "The Census, Social Structure and Objectification in South Asia"; Appadurai, "Number in the Colonial Imagination." Dirks, *Castes of Mind.*
7. Bernard Cohn, *Colonialism and Its Forms of Knowledge* (Princeton, NJ: Princeton University Press, 1996).
8. Nicholas Dirks, foreword to *Colonialism and Its Forms of Knowledge*, by Bernard Cohn, ix–xvii (Princeton, NJ: Princeton University Press, 1996).
9. Cohn, *Colonialism and Its Forms of Knowledge*, 8.
10. By the mid 1850's, the colonial state had begun to plan for an all-India census in 1861. However, the Mutiny both delayed this effort (to quell the potential for further unrest) and confirmed the need for better data on colonial subjects. Dirks, *Castes of Mind*, 200.
11. Cohn, *Colonialism and Its Forms of Knowledge*, 4.
12. Cohn, "The Census, Social Structure and Objectification in South Asia," 242–43.
13. Cohn, *Colonialism and Its Forms of Knowledge*, 8.
14. Dirks, *Castes of Mind*, 199–200.
15. Cohn, *Colonialism and Its Forms of Knowledge*, 8.
16. Censuses occurring in England during this period asked about place of birth. There were no questions about religion, race, or caste in England for the entire period that colonial censuses in India collected these data.
17. Cohn, *Colonialism and Its Forms of Knowledge*, 3–4.
18. The Imperial Census of 1872 census was both incomplete—it failed to cover all territories controlled by the British—and non-synchronic, as data from several provinces were collected during a three-to-six-year window. Although a uniform schedule was used in most places, it was not centrally supervised. Henry Waterfield, *Memorandum on the Census of British India of 1871–72* (London: H.M. Stationery Office, 1875).
19. W. C. Plowden, *Report on the Census of British India Taken on the 17th February 1881* (London: Eyre and Spottiswoode, 1883).
20. Espeland and Stevens, "Commensuration as a Social Process," 318.
21. J. A. Baines, "Conference of Census Officers of the 1881 Census to Consider the Matter Relating to 1891 Census (Feb 10th, 1890)," in *Indian Census through a Hundred Years*, by Dandapani Natarajan, Census of India 1971 (Delhi: Office of the Registrar General, India, 1972), 246.
22. Cohn, "The Census, Social Structure and Objectification in South Asia," 243–44.
23. J. A. Baines, "Letter to George Hamilton (April 29, 1899)," in *Indian Census through a Hundred Years*, by Dandapani Natarajan, Census of India 1971 (Delhi: Office of the Registrar General, India, 1972), 249.

24. Rajendra Lal Mitra, 1881, as quoted in Cohn "The Census, Social Structure and Objectification in South Asia," 245.
25. Dirks, *Castes of Mind*.
26. Ibid.
27. Geetha and Rajadurai, *Towards a Non-Brahman Millennium*.
28. Ibid.; Geetha and Rajadurai, "Dalits and Non-Brahman Consciousness in Colonial Tamil Nadu."
29. Ibid.
30. Pritam Singh, "Numbers as a Means to Power: Politics of Caste as a Census Category in Colonial India c 1871–1941," *Economic and Political Weekly* 57, no. 18 (April 30, 2022): 63–68.
31. Kenneth W. Jones, "Religious Identity and the Indian Census," in *The Census in British India*, ed. Norman Gerald Barrier, 73–101 (New Delhi: Manohar, 1981).
32. Ibid.
33. Cohn, "The Census, Social Structure and Objectification in South Asia," 250.
34. Several of the largest caste associations in contemporary Karnataka, for example, started in the first decade of the twentieth century.
35. Cohn, "The Census, Social Structure and Objectification in South Asia," 243.
36. See note 4 in Chapter 1 for more details on systems of classifying caste.
37. Cohn, "The Census, Social Structure and Objectification in South Asia," 245.
38. Dirks, *Castes of Mind*.
39. Padmanabh Samarendra, "Census in Colonial India and the Birth of Caste," *Economic and Political Weekly* 46, no. 33 (2011): 51–58.
40. Dirks, *Castes of Mind*, 206.
41. Galanter, *Competing Equalities*, 25.
42. Ibid.
43. J. T. Marten, *Census of India 1921, India, Vol. 1, Part II: Tables* (Calcutta: Superintendent Government Printing, 1923), 151.
44. Natarajan, *Indian Census through a Hundred Years*, 186.
45. Ibid., 513–625.
46. The enumerator instructions include subsequent examples of caste for Muslims (along with Hindus, as expected), while continuing to associate Christians with race: "As Brahman, Rajput, Bania, Kunbi for Hindus; Pathan, Moghal, etc. for Mussulmans; Eurasian or Nativ[e] Christians for Christians." Ibid., 537–38.
47. Ibid., 567.
48. Ibid., 606.
49. Ibid., 625.
50. Samarendra, "Census in Colonial India and the Birth of Caste."

51. Ibid., 57.
52. Curtis, *The Politics of Population*, 4.
53. Ghurye, *Caste and Race in India*; Srinivas, *Caste in Modern India and Other Essays*; Cohn, "The Census, Social Structure and Objectification in South Asia"; Appadurai, "Number in the Colonial Imagination"; Dirks, *Castes of Mind.*
54. Cohn, "The Census, Social Structure and Objectification in South Asia," 242.
55. Michel Foucault, *Discipline and Punish* (New York: Vintage Books, 1979); Michel Foucault, "Governmentality," in *The Foucault Effect: Studies in Governmentality*, ed. Graham Burchell, Colin Gordon, and Peter Miller, 87–104 (Chicago: University of Chicago Press, 1991); Michel Foucault, "Power and Strategies," in *Power/Knowledge: Selected Interviews and Other Writings, 1972--1977*, ed. Colin Gordon, 134–45 (New York: Pantheon Books, 1980); Pierre Bourdieu, *Practical Reason: On the Theory of Action* (Stanford, CA: Stanford University Press, 1998).
56. For example, with regard to US trained scholars like Cohn, Eleanor Zelliot completed a Ph.D. in history in 1969. She pioneered scholarship on Ambedkar and Dalit movements and throughout her life introduced Ambedkar and Dalit movements to scholars in the west. Although I studied geology at the time, I was blessed to take an undergraduate Indian history course taught by Professor Zelliot at Carleton College before her retirement in 1997. Eugene F. Irschick completed a Ph.D. at the University of Chicago and published *Politics and Social Conflict in South India* in 1969. Gail Omvedt completed her Ph.D. in sociology in 1973, and her first book, *Cultural Revolt in a Colonial Society: The Non Brahman Movement in Western India*, was published in 1976. Marc Juergensmeyer completed his Ph.D. in political science in 1974, and his research on the Ad Dharm movement subsequently led to the publication of *Religion as Social Vision: The Movement against Untouchability in 20th Century Punjab*.
57. For example, Iyothee Thass contributed to the founding of the *Dravida Pandian* magazine in 1896 and the weekly magazine *Oru Paisa Tamilan* in 1907 (which was renamed to *Tamilan* one year later), which included a circulation among the Tamil working class domestically and the diaspora that migrated to southern and eastern Africa and Southeast Asia. These magazines offered an ongoing critique of Brahmanism and nationalist movements, and highlighted the contributions of Dalits and other caste-oppressed groups. Aloysius, *Nationalism without a Nation in India*; J. Bala, "Pandit Iyothee Thass and the Revival of Tamil Buddhism," Round Table India for an Informed Ambedkar Age, June 27, 2014, https://www.roundtableindia.co.in/pandit-iyothee-thass-and-the-revival-of-tamil-buddhism/, accessed April 1, 2024; Gajendran Ayyathurai, "Colonialism, Caste, and Gender: The Emergence of Critical Caste Feminism in Modern

South India," *Journal of Women's History* 33, no. 3 (Fall 2021): 133–56; Dickens Leonard, "Caste-Less Tamils and Early Print Public Sphere: Remembering Iyothee Thass (1845–914)," *South Asia Research* 41, no. 3 (2021): 349–68.

58. Phule, "Slavery (1873)," 32.
59. Ibid.
60. Omvedt, *Buddhism in India: Challenging Brahmanism and Caste.*
61. Ibid., 237.
62. Phule, "Slavery (1873)," 45.
63. In Pune in the Bombay Presidency, Phule and his wife, Savitribai Phule, opened their first school for oppressed-caste girls in 1848 and devoted their lives to the cause. Similarly, in the Madras Presidency, Iyothee Thass co-founded the Dravida Mahajana Sabha with Rettamalai Srinivasan to push for educational access and broader access to employment for oppressed castes. The organization's 1891 Charter of Demands pushed to diversify local government offices and create guaranteed jobs for Dalits by creating educational access through schools in every village for caste-oppressed children, scholarships for the strongest students from caste-oppressed communities, guaranteed government employment for the students if they pass the secondary school exam, and removal of any barriers to government employment. Paik, *Dalit Women's Education in Modern India*, 76, 150; Geetha and Rajadurai, "Dalits and Non-Brahman Consciousness in Colonial Tamil Nadu," 2091–92; G. Aloysius, *Religion as Emancipatory Identity: A Buddhist Movement among the Tamils under Colonialism* (New Delhi: New Age International, 1998).
64. Paik, *Dalit Women's Education in Modern India*, 80.
65. Salve (1855) cited in Paik, *Dalit Women's Education in Modern India: Double Discrimination*, 150.
66. In fact, to create educational access for oppressed castes and girls from a variety of backgrounds, Phule and Savitribai Phule created and ran schools for these groups.
67. Phule, "Slavery (1873)," 35.
68. Ambedkar, "States and Minorities."
69. Ambedkar, "Thoughts on Linguistic States."
70. Ibid., 33.
71. Rupa Viswanath, *The Pariah Problem* (New York, NY: Columbia University Press, 2014); P. S. Mohan, *Modernity of Slavery* (New Delhi: Oxford University Press, 2015); G. Prakash, *Bonded Histories* (Cambridge: Cambridge University Press, 1990).
72. Phule, "Slavery (1873)," 91.
73. Ambedkar, "Annihilation of Caste," 242–43. Ambedkar says that "in every Hindu the consciousness that exists is the consciousness of his caste."
74. Paik, *Dalit Women's Education in Modern India*, 17–18.
75. Ibid., 74.

76. George Chesney, *Indian Polity: A View of the System of Administration in India* (London: Longmans, Green and Co., 1894), 195.
77. Ibid.
78. The 1901 Census of India enumerated 62 million people as Muslim (that is, 21 percent of the population) and nearly 15 million people as Brahmans (that is, 5 percent of the population) of a total population of over 294 million people. H. H. Risley and E. A. Gait, *Census of India 1901, Vol 1A India, Part II. Tables* (Calcutta: Office of Superintendent of Government Printing, India, 1903).
79. Chesney, *Indian Polity*, 195.
80. Government of Mysore, *Report of the Sir Leslie Miller Committee* (1919).
81. Phule, "Slavery (1873)," 92.
82. Unpublished summary table of Karnataka's history of reservation. Attained from Karnataka's Backward Commissions Office in July 2011.
83. Phule, "Slavery (1873)," 91.
84. Ibid, 83.
85. Ibid.
86. Ambedkar, "States and Minorities."
87. Ibid, 82.
88. Ibid.
89. Ibid.
90. In Phule's accounts English colonizers maybe ignorant or unknowingly complicit, but rarely the cause of the problem.
91. Ambedkar, "States and Minorities."
92. Geetha and Rajadurai, "Dalits and Non-Brahman Consciousness in Colonial Tamil Nadu," 2091–92.
93. Geetha and Rajadurai, *Towards a Non-Brahman Millennium*, 62–70.
94. Omvedt, *Buddhism in India*, 223–224.
95. Omvedt, *Dalits and the Democratic Revolution*, 106.
96. Jaffrelot summarized the long-standing diverging view over caste and social change between Gandhi and Ambedkar. Ambedkar was trying to dismantle the caste system rapidly through conversion to Buddhism, inter-caste marriage, inter-caste dining, and separate electorates to create an organized mass movement against untouchability, led by "untouchables." In contrast, Gandhi's paternalistic view was to protect the "Hindu family," and he stressed the unity of all Hindus and pushed for gradual social reforms at the expense of equality for untouchables in the immediate period. Gandhi argued against untouchability by asserting that "untouchables" should be brought into the *varna* system as Shudras. Jaffrelot, *India's Silent Revolution*, 22–25.
97. Ambedkar, "States and Minorities," 414.
98. Ibid.
99. Ibid.

100. Ibid.
101. As quoted in Natarajan, *Indian Census through a Hundred Years*, 266.
102. Jawaharlal Nehru, as quoted in L. M. Shrikant, *Report of the Commissioner of Scheduled Castes and Scheduled Tribes* (New Delhi: Manager, Government of India Press, 1956), 226.
103. S. Deshpande, "Caste and Castelessness."
104. Subramanian, *The Caste of Merit*.
105. Steven Ian Wilkinson, "India, Consociational Theory, and Ethnic Violence," *Asian Survey* 40, no. 5 (2000): 767–91.
106. Ibid.
107. "Constituent Assembly Debates, Official Report" (New Delhi: Manager of Publications, May 25, 1949), as cited in Wilkinson, "India, Consociational Theory, and Ethnic Violence."
108. *Times of India*, September 17, 1949, as cited in Wilkinson, "India, Consociational Theory, and Ethnic Violence."
109. *Times of India*, December 5, 1949, as cited in Wilkinson, "India, Consociational Theory, and Ethnic Violence."
110. Wilkinson, "India, Consociational Theory, and Ethnic Violence."
111. The Home Ministry is powerful within the Indian bureaucracy, and the home minister is often seen as the most powerful minister, second only to the prime minister.
112. In 1932, the colonial government created the term "Scheduled Castes" to denote those historically outcaste and untouchable groups who would enjoy special electoral arrangements. The 1950 Indian Constitution had codified and extended the system of reservations that developed during the colonial period, in which a proportion of government jobs was reserved for "backward" groups such as SCs and STs—equal to their proportion in the population. The term "backward" appears in three articles of the Indian Constitution. Article 15 states that "nothing shall prevent the State for making any provision for the reservation of appointments or posts in favor of any backward class of citizens which, in the opinion of the State, is not adequately represented, in the Services under the State." *Constitution of India as Amended by Constitution Act 2009* (Delhi: Kamal Publishers, 2011), 6.
113. Caste was for the most part absent from the all-India report of the 1951 census and in the main reports for many states or regions. In most instances the second half of the state-level reports that included tables had a one-page table for SCs and STs. The notable exception was the Madras & Coorg report, which included a chapter on "Backward Communities" in its main report and an appendix that lists the Backward Classes, Non-Backward Classes, Scheduled Castes, and Scheduled Tribes.
114. Lee, *Deceptive Majority*, 69–71; Viswanath, *The Pariah Problem*.
115. Lee, *Deceptive Majority*, 33.

116. Arvind Kumar, "Exclusion of Pasmanda Muslims and Dalit Christians from the Scheduled Caste Quota," *South Asia Research* 43 (2023): 192–209.
117. Ibid.
118. Galanter, *Competing Equalities*, 173.
119. Shrikant, *Report of the Commissioner of Scheduled Castes and Scheduled Tribes*, 219.
120. While the chair was Brahman, several other members of the eleven-person committee were from "lower"-caste backgrounds. The questionnaire sent to state governments consisted of twenty-four areas of feedback, including criteria by which to define "backwardness"; the population, geographic locations, occupations (with individual subsections on government service and agriculture), and socio-economic, health and educational status of backward classes within the state; organizations working to bring about "political consciousness" and state agencies "working for the uplift" of backward classes; and possible revisions to the lists of SCs and STs. Shrikant, *Report of the Commissioner of Scheduled Castes and Scheduled Tribes*, 228, 240.
121. The commission asks state governments the following question: "Do you think that the abolition of the mention of caste or sub-caste in the census of 1951 has been useful or otherwise for determining the condition of backward classes? What procedure would you recommend for adoption in the future censuses?" Shrikant, *Report of the Commissioner of Scheduled Castes and Scheduled Tribes*, 226.
122. Backward Classes Commission (K. Kalelkar, Chair), *Report of the Backward Classes Commission*, 11.
123. Ibid, 159.
124. It used the following criteria for these classes, including 837 as the "most backward": (*a*) low social position in the traditional caste hierarchy of Hindu society; (*b*) lack of general educational advancement among the major section of a caste or community; (*c*) inadequate or no representation in government services; (*d*) inadequate representation in the field of trade, commerce, and industry.
125. Susan Bayly, *Caste, Society and Politics in India from the Eighteenth Century to the Modern Age*, The New Cambridge History of India (Cambridge: Cambridge University Press, 1999), 291.
126. Galanter, *Competing Equalities*, 173.
127. He argued, "[I]t would have been better if we could determine the criteria of backwardness on principles other than caste," and that the "caste test was repugnant to democracy." Backward Classes Commission (K. Kalelkar, Chair), *Report of the Backward Classes Commission*, vi–xiii.
128. The minister of home affairs, in a strong critique that accompanied the report when it was put before parliament in September 1956, argued that the Kalelkar Commission had failed to find "positive and workable criteria" and recommended that state governments "give all reasonable facilities

… in accordance with their existing lists and also to such others who in their opinion deserve to be considered as socially and educationally backward in the existing circumstances," and undertake occasional surveys to identify backward classes.

129. For example, by the 1950s, the provinces and princely states found in present-day Kerala, Tamil Nadu, Andhra Pradesh, Karnataka, and Maharashtra already had reservation benefits in place for OBCs, for recruitment into educational institutions and government jobs. The definition of OBC varied across these provinces; colonial-period committees had helped to define "depressed," "backward," and "non-Brahman" classes in particular regions, beginning with the Millers Committee in Mysore (present-day Karnataka) in 1918 and continuing into the 1920s in several other states in the south and Bombay Province (part of present-day Maharashtra). At the same time, limited educational concessions were made for OBCs in Bihar and Uttar Pradesh beginning in the late 1940s. Galanter, *Competing Equalities*, 155–68.
130. Ibid., 172–73.
131. Backward Classes Commission (K. Kalelkar, Chair), *Report of the Backward Classes Commission*, 4.
132. In the case of present-day Karnataka (previously known as Mysore), reservations had existed in the region since 1921. The first post-colonial Backward Classes Commission, headed by R. Nagana Gowda, completed its report in 1960. Unlike the 1919 Miller Commission report, the Nagana Gowda Commission report's definition of Backward Classes did not include SCs and STs, who had their separate central government lists and reservation quotas in central and state government institutions. Under Article 15(4) of the Constitution, the Nagana Gowda Commission recommended a reservation in educational institutions for groups found on both the socially and educationally backward lists. This population accounted for approximately 57 percent of the population and had 22 percent reservation in educational institutions. SCs and STs had a reservation quota equal to their proportion in the state population: 15 percent and 3 percent, respectively. The commission also recommended that "all those communities which are backward and are not adequately represented in Government Services should be considered as the 'Other Backward Classes' in the State for the purpose of reservation in posts in Government Service under Article 16(4) of the Constitution." A 25 percent reservation was set aside for OBCs in state government jobs, and again the SC and ST quotas were equal to their proportions in the state population. Government of Mysore, *Mysore Backward Classes Committee Final Report (Nagana Gowda Committee)* (Bangalore: Government Press, 1961), 22.
133. Interview with government official, July 19, 2011.
134. Galanter, *Competing Equalities*, 163.
135. Ibid., 168.

136. Ibid.
137. Ibid., 181–82.
138. In the case of present-day Karnataka, three such legal cases, *Ramakrishna Singh v. State of Mysore*, *Partha vs. State of Mysore*, and *Balaji vs. State of Mysore*, put ongoing scrutiny on how OBCs should be defined and whether the state's classification of the category along caste lines, versus defined groups along other identities, was acceptable.
139. The "Emergency" during prime minister Indira Gandhi's time in power lasted for twenty-one months between 1975 and 1977 and led to the suspension of elections and civil liberties and was justified by the prime minister.
140. Backward Classes Commission (B. P. Mandal, Chair), *Report of the Backward Classes Commission*, iv.
141. However, several states had reservation quotas that were higher than 50 percent.
142. Ibid.
143. The order (Memorandum No. 36012/31/90-Estt. (SCT) dated 13th August 1990) specified that Socially and Educationally Backward Classes eligible in the first phase would include the castes and communities which are common to both the lists in the report of the Mandal Commission and the state governments.
144. Writ Petition (Civil) No.930 of 1990 – *Indira Sawhney and Others vs. Union of India and Others.*
145. A. Ramaiah, "Identifying Other Backward Classes," *Economic and Political Weekly* 27, no. 23 (June 6, 1992): 1203–07, 1207.
146. See M. *Nagraj & Ors v. Union of India* and *Suraj Bhan Meena v. State of Rajasthan.*
147. Interview, May 9, 2013.
148. M. Vijayanunni, "Caste Can't Be Caste Off," *The Indian Express*, November 19, 1999.
149. The powerful Home Ministry houses the ORGI along with border security, the centralized police services, internal security, disaster management, and other security and terrorism units.
150. Gandhi, "India Round Table Conference (Second Session),"159. M. K. Gandhi, "Harijan (5 January 1934)," in *Collected Works of Mahatma Gandhi (8 October 1933–17 January 1934)*, vol. 62 (New Delhi: Publications Division Government of India, 1999), http://www.gandhiashramsevagram.org/gandhi-literature/mahatma-gandhi-collected-works-volume-62.pdf, accessed April 1, 2024.

3

The Politics of the Count*

> The Census, perfect though it is not, is the most competent, in fact, the only competent agency that can be expected to undertake the all-India data collection and tabulation exercise required for caste data.
>
> —M. Vijayanunni, 2010[1]

M. Vijayanunni, who was now retired from the Indian Administrative Service, continued to advocate for a full caste-wise enumeration in the census in the lead-up to Census 2011. As the former head of the ORGI, or the Indian census bureau, Vijayanunni spoke from a unique position when he argued that the ORGI was the only agency with the technical experience to undertake the task. Vijayanunni's certainty stemmed from the technology available for cleaning, analyzing, and tabulating data and his detailed historical knowledge of censuses. If the colonial state could manually process, tabulate, and publish large volumes of caste and religion data, the ORGI was capable of completing a similar task in 2011. Vijayanunni argued that the question was not whether the ORGI had the *capacity* to do so, but whether it was *willing* to do so. Politics was central in deciding the path forward.

A coalition of organizations, activists, politicians, and public intellectuals—including Vijayanunni—came together to target a change in census policy on caste. This chapter describes the advocacy efforts to include a caste count in Census 2011, and then details the institutional backlash that followed the campaign's success. In examining the steps that executive bureaucrats and

* An earlier version of this chapter was published as "Counting Caste: Census, Politics and Castelessness in India," *Politics and Society* 46, no. 4 (2018): 455–84, and is reprinted with permission. Parts of this chapter overlap with material in a co-authored publication with Diana Graizbord and Cedric de Leon, "The Retreat to Method: The Aftermath of Elite Concession to Civil Society in India and Mexico," *Studies in Comparative International Development* 54 (2019): 19–39, and are reprinted with permission.

allied groups took to keep a caste-wise enumeration out of Census 2011, this chapter traces the strategies of bureaucratic deflection and how they reinforce castelessness in the census, the state, and beyond.[2]

Securing the Concession: Caste Census to Document Systemic Inequality

Dilip Mandal had worked for several media houses as a writer, editor, and producer for more than fifteen years. His most recent position as managing editor for a national magazine had thrust him into the limelight. Mandal had broken a "glass-ceiling" by being one of a handful of journalists from an oppressed-caste background to enter the editorial ranks, yet his road to becoming a senior journalist had not been easy. He spoke out against caste-based discrimination in India's mainstream media and other major institutions of society and in 2009 joined forces with other public intellectuals, activists, and politicians to push for a change in census policy on caste. In a 2010 interview, Mandal describes how "the people who oppose caste-census are very less in number, but they are very powerful, and they run this country."[3] Mandal named the elite yet powerful nature of the individuals and groups working behind the scenes to keep a caste-wise enumeration out of the census.

Much of the early organizing in support of a caste-wise enumeration in Census 2011 came from groups that desired caste-wise data for the implementation of OBC reservations. In October 2008, a leader of the Pattali Makkal Katchi (PMK) party spoke in support of a caste-wise enumeration in the Lok Sabha. Member of parliament (MP) P. Ramadass, representing the PMK in Pondicherry, described a common complaint made by judges and advocates during legal proceedings related to the extension of affirmative action:

> Recently when the Parliament passed the 27 percent reservation, it was stayed by the Supreme Court. One of the reasons given by the Supreme Court was that we are giving 27 percent reservation to OBC students in higher educational institutions but where is the data of the population of the OBC? On what basis you are fixing this 27 percent? The Government of India could not provide a ready answer. They said that we rely upon the census of 1931 and we rely upon the Mandal Commission Report, which extrapolates the population of 1931. But Supreme Court was not satisfied. Therefore, there is an immediate need, imperative need that the Government of India should go in for caste-census from the impending census operations.[4]

Ramadass highlighted the need for updated census data on OBCs to implement reservation policy. During these early advocacy efforts, the issue was commonly framed as a particularistic one; caste-wise data were needed for the purposes of administering OBC reservations and other government programs in support of OBCs, similar to the ORGI's collection of data on SCs and STs in the decennial census.

The next day, Ramadass joined a delegation of MPs led by the union health minister to meet the home minister and demand a caste-wise enumeration in Census 2011. The delegation delivered a memo signed by 170 MPs from diverse political parties. Many in the delegation had supported the expansion of OBC reservations and saw this policy change as being necessary to implement reservation policy.[5] Others saw the need for caste data as necessary to develop a broader set of anti-caste and pro-poor policies. Yet the home minister was unmoved by their request and remained unsupportive of the policy change.

A handful of parliamentarians in the Rajya Sabha, or the upper house of parliament, also started to make demands to include a caste-wise enumeration in Census 2011. On December 9, 2009, MP Matilal Sarkar from Tripura made a statement about the importance of enumerating OBCs. Sarkar emphasized the need to determine the size of the OBC population, similar to how "the determination of facts about SC and ST comes under Census program."[6] About a week later, MP Ali Anwar Ansari also made a statement in the Rajya Sabha arguing for a caste-wise enumeration in the upcoming decennial census to ensure that reliable data on the social, economic, and educational status of castes informed government schemes and programs.[7] Ansari's position represented the view of many in the Pasmanda Muslim movement, which had brought attention to the stark variations in life opportunities and experiences based on caste and class within the Muslim community. The census enumeration of caste would reveal the discrimination and structural inequalities that Dalit and other caste-oppressed Muslims experienced. While some states had reservations for Muslims as part of their state-level OBC quotas, reservations for Muslims in the central government remained disputed despite the 2006 expert committee report headed by retired justice Rajindar Sachar that documented the systemic discrimination and inequality of opportunities that Muslim communities experienced in India.[8]

The minister of state for home affairs, Ajay Maken, responded to Ansari in a letter dated February 20, 2010, summarizing the political leadership's position. Maken explained that the "decision to discourage community distinction based on caste was taken keeping in view the spirit of the secular state enshrined in the Preamble of the Constitution of India." Maken argued that the enumeration of OBCs would be very difficult as "the criteria for enlisting OBCs is not uniform but varies from State to State and from State to the Centre."The process of classifying OBCs would "make the Census Operation a cumbersome process" and "may jeopardize the very process of Census Operation."While Maken acknowledged the need for data on OBCs, he simultaneously argued that the decennial census should be protected from the collection of caste-wise data:

> While not denying the fact that there is a necessity to have the population details of OBCs in the country for better formulation of policy, it is felt that the population census is not the ideal instrument for collection of these details. The operational difficulties are so many that there is a grave danger that the basic integrity of the Census data may be compromised, and the fundamental population count itself could get distorted.[9]

Maken's concern for the "basic integrity" of census data was a position that would be repeated by the home minister, senior census officials, and Congress leadership in the year to follow.[10] In prioritizing the need to protect the "integrity" of census data, the political leadership silenced debates over ideological differences and elevated technical considerations. They used the concept of integrity with its multiple valences narrowly. Integrity has technical and ethical meanings. All parties respected the census for its technical strength and history of producing reliable data. Built into their use of "integrity" was the assertion that including a caste-wise enumeration in Census 2011 would damage the technical aspects of the project. Yet no evidence existed that the technically strong aspects of the census would be harmed by the caste-wise enumeration, although Maken's language of "grave danger" implied otherwise. Caste census advocates, however, challenged the ethics of not collecting caste-wise data. Caste-wise data were crucial to the implementation of reservation policy and for other social justice efforts. Maken and the political leadership sought to minimize scrutiny and debate on the ethical aspects of integrity when they cited operational barriers to the collection of caste-wise data. They sidelined a conversation over ethical considerations when they reframed "integrity" as solely a technical consideration. MPs Ramadass, Sarkar, and Ansari had raised ethical and policymaking considerations. While Maken formally replied to Ansari's letter, the ORGI undoubtedly provided the language to describe the technical challenges of collecting caste-wise data in

Census 2011 in order to block efforts to include a caste-wise enumeration in the census.

A small group of scholars in support of a caste-wise enumeration in the census also tried to influence census policy by meeting with top political leaders. In late 2008 and early 2009, Satish Deshpande and Yogendra Yadav, two public intellectuals based in Delhi, wrote a memo that argued in favor of a caste enumeration in Census 2011 and circulated it among the Congress party leaders. Their efforts resulted in a meeting with Rahul Gandhi, who was then Congress Party's general secretary and an MP. Based on this meeting and other conversations with the political leadership, Yadav and Deshpande learned that, like the BJP-led government in the lead-up to Census 2001, Congress had little interest in collecting caste-wise data in the census.[11] Deshpande and Yadav later joined forces with other scholars, public intellectuals, and activists advocating for a caste census in late 2009 and 2010.

Groups advocating for a caste-wise enumeration in Census 2011 also filed petitions through the courts. The judiciary had repeatedly asked for updated caste-wise data during cases that related to the extension of reservations, as referenced by MP Ramadass. The question from the court was usually a variation of "how is it possible to make a determination about expanding reservations to group X, when we lack reliable data about group X?" While all levels of the judiciary asked for credible data on caste, the Supreme Court generally refused to interfere with census policy on caste, while state-level high courts (particularly in the south) supported petitions that advocated for the collection of caste-wise data in the census.

In one such case, S. Ramadoss, the founder of the PMK party, filed a legal petition in 2009 to include a full caste count in Census 2011 as part of long-standing efforts to increase the Vanniyar community's reservation within the OBC quota in Tamil Nadu.[12] The Supreme Court declined to decide on the petition, and the chief justice requested that the PMK withdraw the petition, stating, "[W]e can't give any direction in matters of policy. This [caste-based census] may cause immense caste strife. It has serious implications. That is why it has not been done for the last 60 years...."[13] The chief justice's representation of a caste census as causing "immense caste strife" likely alluded to the history of oppressor-caste-initiated violence in regions of north India following the expansion of affirmative action benefits in the 1990s and 2000s. While the chief justice described a change in census policy as potentially dangerous, his statement also indirectly implied that the status quo in census policy was not dangerous.[14] The Supreme Court remained "non-interventionist" and chose not to decide on petitions advocating for a caste-wise enumeration in Census 2011.

In contrast, advocacy organizations found support in the lower courts. The Madras High Court continued to consider these types of petitions and sought direction from the ORGI after reviewing a petition by the Advocates Forum for Social Justice (AFSJ).[15] Similar legal petitions came out of other southern states during this period, and their respective high courts supported the petitions.[16] The ORGI responded to the AFSJ's petition by stating that it could not include any question related to caste in Census 2011 except for the enumeration of SCs and STs. It also referred the policy decision regarding the enumeration of caste to the Ministry of Social Justice and Empowerment. Less than two months later, in May 2010, the Madras High Court passed an order in support of a similar petition by R. Krishnamurthy and directed the census commissioner to take measures to conduct a caste-wise census.[17] In doing so the High Court echoed previous sentiments by the Supreme Court that projections of OBC population size based on 1931 data were inadequate and recent data were necessary. This did not lead to a change in policy, however.[18] These early exchanges between the lower courts and the ORGI, and parliamentarians and the executive bureaucracy, reveal the central bureaucracy's resistance and the strategies that would be fine-tuned and later employed to keep a caste-wise enumeration out of Census 2011. A more coordinated campaign to change census policy on caste began in late 2009 that built upon these earlier efforts through parliamentary inquiry, private lobbying, and the courts.

A Delhi-based civil society organization named Janhit Abhiyan started to organize MPs and collaborated with other organizations and activists in support of a caste-wise census. The nongovernmental organization (NGO) already had legitimacy in civil society as it had been involved in several campaigns related to the implementation of OBC reservations.[19] A journalist involved in the caste census campaign explained:

> Janhit Abhiyan did not just pop up for the caste census; if they did, they wouldn't have been able to garner support from a wide constituency. They wouldn't have been taken seriously. They had been involved in several campaigns related to the implementation of reservations and increasing the caste quota.[20]

Over the next several months Janhit Abhiyan began to meet regularly with other organizations in Delhi to mobilize a broader constituency in support of an expanded caste count in Census 2011. Janhit Abhiyan co-sponsored public forums to discuss the enumeration of caste and garnered the backing of organizations that represented the interests of Dalits and OBCs. In December 2009, Janhit Abhiyan organized a conference attended by public intellectuals, journalists, MPs, and the general public. Raj Narayan, the convener of Janhit Abhiyan, argued:

> [A] caste-wise census is necessary for collecting reliable and accurate data relating to economic, social, educational, and political status of different castes. Since no such exercise has been taken in independent India, the fruits of development and benefits of the government schemes are not able to reach the targeted sections. It is not a question of collecting data about the status of only backward castes but also relating to economically poorer sections among the upper castes. The government can redesign its welfare schemes based on the latest data.[21]

Narayan and others collaborating with Janhit Abhiyan actively worked to shift the language of the debate from an issue important to OBCs to one of exclusion more broadly. Narayan believed, "[I]f a caste-wise census is done properly it will help to eradicate the caste system in this country. The concept was not to empower caste but to eradicate caste."[22]

The shift in language reflected a widening of support for a caste-wise enumeration in the census. By early 2010, Janhit Abhiyan had expanded the campaign to include key government employee organizations: the All India Backward and Minority Communities Employees Federation (BAMCEF) and the Backward Classes Federation in the north Indian state of Uttar Pradesh.[23] Janhit Abhiyan also collaborated with MPs who were staunch advocates for a caste census to help educate other parliamentarians about the issue. They first sent a packet of information about the need for a caste-wise enumeration in the census to every parliamentarian, and then visited the offices of key MPs in March and April 2010 to push for a united front across party lines.[24] MP Sharad Yadav of the Janata Dal United (JD-U) was a key supporter of the issue and repeatedly spoke in favor of a caste census. Yadav and other MPs from OBC communities formed a new political class that had challenged Brahman and oppressor-caste control of politics in the northern Hindi belt.[25] Yadav saw the government's refusal to include a caste-wise enumeration in the census as part of a global strategy to maintain caste power:

> Most of the Indians who went abroad are upper-caste, so the rest of the world does not know about caste because the upper-caste pretends it doesn't exist.

> But, in reality, caste is worse than Apartheid, it is worse than White/Black relations in the US. In this country, it is not religion or language, but only caste that is universal from Kerala to Kashmir. You can change your religion, but you can't change your caste. That is why there are Dalit Christians and Dalit Muslims.... In the lead up to the 2001 census, during the first 15 days in office, the JD-led government of Devagowda passed a resolution/order to conduct a caste census. But it has still not been implemented because those in office since then [that is, Congress and BJP] are controlled by upper castes.[26]

Yadav highlights the domestic and global barriers to challenging caste. He also describes the pervasiveness of the ideology of castelessness, which obscures the continuing power of caste elites and the intractability of caste. Over the course of several months of intensive organizing, MPs such as Sharad Yadav and Mulayam Singh Yadav of the Samajwadi Party (SP) worked closely with Raj Narayan, Dilip Mandal, Satish Deshpande, M. Vijayanunni, and other well-known public intellectuals and activists.[27] Advocates for an expanded caste count hoped to build a base of support while shifting the discourse from a particularistic issue concerning OBCs to a social justice issue. They sought to organize a diverse coalition in support of change.

Widespread Political Support to Change Census Policy on Caste

The organizing efforts for a caste census culminated on the floor of the Lok Sabha in early May 2010, as MPs sought to put political pressure on the Congress-led coalition government and force the cabinet to change its position. The start of the parliamentary debate also aligned with the beginning of the household listing phase of Census 2011.[28] With the first stage of the census underway, the topic of the caste-wise enumeration was of high importance. MPs deliberated the issue as part of a "discussion under rule 193"—which meant there would be a debate on the topic but no official vote.

The Lok Sabha debate was lively, spread across several afternoons, and nearly all the MPs who spoke were in favor of a caste-wise enumeration in the census. Several recurring themes emerged during the debate. First, the MPs agreed that caste continued to be an axis of inequality in contemporary India. An excerpt from MP Gurudas Dasgupta's speech delivered on the floor of the Lok Sabha on May 6, 2010, summarizes this common view:

> Mr. Chairman, I join unhesitatingly the unanimity of the House that the identification of a person should be on the basis of social status, that is, caste.

> I agree with it totally. The fact of the matter is that the Scheduled Castes, Scheduled Tribes, OBCs, and Muslims are subjected to social atrocity all over the country and nobody can deny it. We have not been able to break poverty; nor have we been able to bring about economic empowerment of the people. It is a reality that the most poor people of this country belong to this section. Who are the most poor people? They are the Dalits, they are the Adivasis, the Scheduled Castes, the Scheduled Tribes, the Muslims and the Tribals. If it is so, then what is wrong in identifying them on the basis of their own identification status, that is, caste?... The social system is based on caste system. Therefore, it should be taken into account while the census is done. This is my submission in unity with the cross-section of the opinion that has been reflected in the House.[29]

In addition, MPs repeatedly made the case that since discrimination occurred along caste and religious lines, government policies needed to work to ameliorate the position of oppressed groups and that the identification and enumeration of these groups were necessary to implement policies and programs. MP Sameer Bhujbal argued this position during his speech:

> More importantly, as long as the Government offers affirmative action for groups that are backward for historical reasons, it must gather as much information about them as possible. It also would help the policymakers in identifying more precisely just which groups are disadvantaged and to what extent. This is essential because policy can then deal with facts rather than impressions, as is the case now in the absence of any authoritative data. In fact, inclusion of "caste" in the census will help allocations to be made for OBCs and other marginalized groups.[30]

The importance of caste-based data for the successful implementation of reservation policy remained the core motivation for a caste-wise enumeration in the census. Building upon comments by MP Mulayam Singh Yadav and MP Dara Singh Chauhan, MP Sharad Yadav presented data on the under-representation of OBCs in the central government. He emphasized that caste-wise data were necessary to inform government policies and programs, yet warned that people committed to social justice should not be deceived into thinking that "this matter will get solved just by counting numbers."[31]

While several speeches emphasized the need to collect data to improve the lives of caste-oppressed groups, MPs also suggested how the data could be collected. MP B. Matab recommended that the collection of caste-wise data should occur during the second stage of the census in February 2011.[32] Several MPs explained that data on OBCs could be collected with minor revisions to the census schedule; the ORGI could simply add the answer option of "OBC" to the existing options of "SC" and "ST," changing the default of "not SC or ST" to "not SC, ST, or OBC."

As part of the chorus of support, the opposition BJP also came out in favor of the caste census. The BJP had not agreed to this policy change when it was in power during Census 2001. Now, the leader of the opposition, MP Sushma Swaraj, argued that a caste-based survey would not aggravate caste-based discrimination. Since caste was already the basis of job applications and admissions to schools and colleges, she reasoned that the government required census data on caste.[33]

In response to the widespread support for enumerating caste in Census 2011, home minister Chidambaram tried to reframe the debate around the primary purpose of the census and technical considerations. He urged the MPs to protect the decennial census from the consequences of adding a caste-wise enumeration:

> Let me reiterate that the main objective of the population census is to do an accurate de facto headcount of the usual residents in India on the deemed date i.e., 00.00 hours on March 1, 2011. Based on universally applied scientific demographic tools, we have an estimate of what the population will be on that day. However, it is necessary and desirable to make an accurate headcount. Hence, the Census. I am sure honorable Members will agree with me when I say that nothing should be done that may affect the accuracy of the headcount or the integrity of the population census.[34]

By suggesting that a caste-wise enumeration might "affect the accuracy of the headcount" or "the integrity" of the census, Chidambaram put forth a technical argument that nonexperts would have difficulty refuting.[35] Nonexperts had minimal legitimacy in participating in a discussion on the factors that influenced the accuracy of census data. Lalu Prasad Yadav of the Rashtriya Janata Dal (RJD), Mulayam Singh Yadav, and Sharad Yadav recognized the strategy and forced an adjournment of the Lok Sabha following the home minister's speech. They held a private meeting with finance minister Pranab Mukherjee, who appeared supportive of their position, and demanded a change in census policy.

In response to the coordinated campaign and support of MPs across party lines, the Congress leadership felt trapped. Congress feared refusing an organized OBC constituency and the potential threat of political parties with an OBC base forming an alliance with the opposition BJP. Congress had criticized the Mandal Commission report in the late 1970s and the extension of reservations to OBCs in the early 1990s, both developments that occurred when non-Congress governments were in power.[36] Congress later worked to maintain its "coalition of extremes" in north India—Brahmans, Scheduled Castes, and Muslims—by promising reservations to low-income caste elites and additional concessions to SCs in the 1990s.[37] A judicial decision forced Congress to adjust its position, including the concession that caste was a defining feature of "backwardness."[38] By the second half of the 1990s, Congress projected itself as a spokesperson for OBCs, although with late and limited success given the dominance of regional political parties that had long supported backward classes in the south and more recently in the north. The Congress leadership did not want to repeat its past mistakes and face electoral consequences. The lack of dissenters among opposition political parties during the discussion made it difficult for Congress to oppose the demands of caste census advocates. The Lok Sabha debate concluded on May 7, 2010, with prime minister Manmohan Singh making a commitment that the cabinet would finalize a decision quickly.[39] Later that same day, finance minister Mukherjee publicly confirmed at a press conference that the political leadership had assented to the policy change.[40] Newspapers throughout the country reported this historic decision on May 8, 2010, including *The Hindu*:

> The government on Friday took the historic decision to include caste in the ongoing census. "Caste will be included in the present census," Finance Minister Pranab Mukherjee told reporters on Friday. A caste-based census was last carried out in 1931. Independent India had shunned counting the numbers of people belonging to each caste—barring an omnibus figure for the scheduled caste—with the objective of moving towards a casteless society.[41]

Caste census advocates were euphoric in their long-awaited victory but also rejoiced in disbelief at their triumph. Their centralized campaign had secured a public concession by top political leaders to change census policy on caste. At the same time, the announced policy change generated an immediate counterresponse from public figures and academics, and the start of less visible institutional strategies to prevent a change in census policy on caste from within the executive bureaucracy.

Public Figures and Campaigns against Caste Census: "My Caste Is Indian"

In his official blog on May 10, 2010, Bollywood superstar Amitabh Bachchan described how municipal workers visited his home during the household listing phase of Census 2011.[42] The interviewers asked him basic questions about the household, including name, place, age, dependents, place of birth, and years of residence. He awaited for them to ask about "the crucial controversial one" but to his surprise the enumerators did not ask the ubiquitously recognizable actor about his caste.[43] Bachchan believed that the census enumerators omitted the question because they were "wary of the sensitivity that this [question] has raised recently by politicians."[44] Undeterred by the omission, the movie star offered up what his answer would be if they had asked the question, "I tell them irrespective of what the system decides, my answer is ready—caste, Indian [*meri jati Hindustani*]."[45]

Unfolding just days after the publicly announced change in decennial census policy, Bachchan's statement spread like wildfire. The media portrayed it as a clever, seemingly progressive, and appropriately nationalistic response. Bachchan's perspective—similar to public intellectuals who dissented to the policy change—complemented the views of Congress Party political leaders who sought to elevate an Indian identity above caste after independence. Opposition to the policy change from well-known individuals ranging from Bollywood stars, such as Bachchan, to public intellectuals received considerable attention.

Journalists and scholars in the Nehruvian tradition argued that the Congress leadership had capitulated to identity politics. Award-winning journalist Barka Dutt criticized the government's acquiescence to a caste-wise enumeration in the census in a newspaper editorial:

> Till just before the Prime Minister indicated his assent to it in Parliament, the Congress had, in fact, been divided on the issue. Pressure from the Yadav allies may have swung the pendulum in a certain direction. But it's a path that could take India back by decades. So, when the census official comes knocking on your door, do what Amitabh Bachchan did. Say: My caste is Indian [meri jati Hindustani]. I wonder whether they have a separate column for that.[46]

Dutt advocated for the Bollywood actor's position as a forward-looking alternative to the caste-wise enumeration, which she argued was a regressive policy. Pratap Bhanu Mehta, a political theorist who was leading a premier research and policy institute in Delhi at the time, described his ideological opposition based on long-standing ideals of democratic India:

> The decision to, in principle, enumerate caste in the Census is a monumental travesty. At one stroke, it trivialises all that modern India has stood for, and condemns it to the tyranny of an insidious kind of identity politics. The call to enumerate caste in the Census is nothing but a raw assertion of power wearing the garb of social justice, an ideological projection of Indian society masquerading under the colour of social science, and a politics of bad faith being projected as a concern for the poor. It is not news that India is deeply structured by hierarchies of various kinds, including caste. These hierarchies still appallingly define structures of opportunity and oppression. But the vision of a just and modern India was founded on an aspiration to promote justice without falling into the same pinched up identities that had kept us narrow and bigoted for so long. The premises of a caste census reproduce the very things we had so long laboured to fight....[47]

Like Dutt, Mehta's opposition to the caste count was explicitly ideological. He argued that a caste-wise enumeration would reinforce identities that modern India should have left behind. He squarely aligned himself with the Nehruvian liberal democratic tradition, like many other public intellectuals.[48] This "nationalism of castelessness" supported the status quo in census policy on caste. In the months to follow, other public intellectuals also weighed in with dissenting opinions on the proposed policy change, describing it as "a monumental travesty" that would "inject the caste poison into India's governance, polity, and society."[49]

Paralleling the Bollywood star's statement, but back in the capital city, senior journalist Ved Pratap Vaidik helped to convene a symposium and start the *meri jati Hindustani* campaign. The campaign sought to have households respond "Indian" during the census enumeration of caste.[50] Vaidik argued that a caste-wise enumeration was anti-nationalist. He instead promoted a casteless Indian identity. The leaders of the *meri jati Hindustani* campaign met with the home minister, held press conferences, conducted a one-day fast, and organized a "People's March" in Delhi.[51] The national media covered the campaign, and a prominent academic described it as an "active campaign by sections of enlightened society."[52]

Yet the *meri jati Hindustani* campaign sidelined the power of being unmarked from the debate. The campaign remained silent on how the most widely respected survey in India could make visible the accrued advantages of caste elites with regard to property, education, employment, and access to public amenities, among other things. This "nationalist" position minimized a more complicated relationship between caste, nationalism, and democracy, which many minoritized groups have experienced. Gopal Guru describes the intellectual and practical struggle that many Dalits and caste-oppressed people face to "comprehend the riddle of nationalism and its ideological frameworks—which makes a rhetorical claim for social equality but sustains the spatial practices of exclusion."[53] Uncritical nationalisms—such as the *meri jati Hindustani* campaign—allowed caste and class elites to easily inhabit a universalist "modern" Indian identity representative of "all," yet largely unavailable to caste-oppressed people. Anti-casteist efforts and movements were critiqued as promoting caste (while they instead challenged an ideology of castelessness), deemed irrelevant and outdated, and given little support to formulate and grow. As such, challenges to caste power are difficult to organize as they are labeled anti-Indian, anti-Hindu, or simply disrespectful and disruptive. Alongside the strengthening of public campaigns against a caste count, central government bureaucrats continued to block a caste-wise enumeration in the decennial census.

Bureaucratic Deflection: Internal Strategies to Block Caste-wise Enumeration in Census

Executive bureaucrats operated in less visible spaces to keep the caste-wise enumeration out of Census 2011. Their efforts highlight how bureaucratic deflection operates, through institutional strategies that emerged relationally in response to organizing by caste census advocates.

Decentralize Decision-making to State Governments

The most common bureaucratic response was to recommend that state governments collect caste-wise data. As stated earlier, the minister of home affairs, Ajay Maken, had argued that the enumeration of OBCs would be difficult given variations in how the category of OBC was defined at the state and central government levels.[54] Asserting that enumerating OBCs "may jeopardize the very process of Census Operation," Maken recommended that state governments should instead collect caste-wise data "since the criteria for determination of OBCs are state specific" and state-level data could be used "for the purpose of reservation and delivering other benefits."[55] Home minister Chidambaram similarly reiterated these points in his statement before the Lok Sabha in May 2010. He said that state governments, via their BCCs, should collect caste-wise data. He emphasized the logistical and technical challenges of enumerating OBCs in the decennial census since both central and state government lists existed.[56] By punting the enumeration of caste to lower tiers of government, Maken, Chidambaram, and other Congress leaders aligned with their predecessors who took a similar approach to OBC reservations in the post-independence period. Yet, unlike in 1955, when the First Backward Classes Commission recommended the collection of caste-wise data in the 1961 Census, the central government had now implemented OBC reservations and therefore required caste-wise data for its own programmatic and policy purposes.

Decentralizing the caste-wise enumeration to state governments would push the project outside of the ORGI's jurisdiction and leave each state to plan, conduct, and carry out the project as they saw fit. Some state governments might conduct a caste-wise survey, while others might not. In addition, state-level units had inconsistent expertise and varied experiences carrying out surveys. The quality of the data would most likely be inconsistent, as with past BPL surveys conducted by state agencies in charge of rural development.

Elevate Technical Concerns

Over time, the executive bureaucracy obscured internal ideological opposition to the caste count. While a range of ideological positions related to the caste census existed, the public conversation shifted away from the ongoing nature of caste-based discrimination to technical considerations.

Just prior to the May 2010 Lok Sabha debate, private differences remained within the union cabinet with regard to census policy on caste. In a cabinet meeting, several ministers argued that caste data should be collected as part of the decennial census. Law minister Veerappa Moily summarized this position later when he said, "[W]hat we want to build is a casteless society. But it cannot be done ignoring reality."[57] In contrast, commerce minister Anand Sharma and social welfare minister Mukul Wasnik spoke out against a caste-wise enumeration. A third group of ministers skirted their ideological position and focused on the logistical challenges of including a caste-wise enumeration in the census. This faction, led by the home minister, argued that the decennial census was not the "ideal instrument" for this task. The home minister expressed that as a social-scientific tool the decennial census should remain separate from the politics of caste.[58] In the year to follow, the public position of the government fell in line with the home minister and the ORGI's views. Ministers who were supportive of a caste-wise enumeration or who opposed a caste census on ideological grounds rarely shared their views publicly after the summer of 2010.

During the Lok Sabha debate, the home minister perfectly summarized the executive bureaucracy's public stance when he refused to state his ideological position. Instead, he argued against a change in census policy on technical and methodological grounds:

> There are two questions here. The first question is, "whether it is desirable to enumerate the caste of each member of the household?" The second question is, assuming that it is desirable to do so, "is the census the vehicle to carry out the enumeration?" I do not wish to enter into a debate on the first question. There can be different views on the subject, and we must respect each other's views.... It is the second question that is relevant for the present discussion.[59]

Chidambaram deemed ideological debate as irrelevant. He instead argued that the conversation should focus on technical and methodological considerations. That is, *what is the best venue to collect caste-wise data?* The home minister depoliticized a highly political discussion and reframed it as technical. The executive bureaucracy utilized this approach regularly and it became a recurring strategy in the year that followed.

The Silencing of Non-experts: "The Retreat to Method"

The executive bureaucracy began to retreat from its public concession at the same time that advocates for a caste census started to celebrate their victory. A three-step processual mechanism unfolded, which I call "the retreat to method" in an article co-authored with sociologists Diana Graizbord and Cedric de Leon.[60] The first step involves a "genuflection to inclusion"—when the state responds favorably to organized demands, in this case from caste census advocates, and in doing so appears legitimately democratic. The prime minister's statement at the end of the Lok Sabha debate and the finance minister's concession to the press the following day complete this step. Next, state actors redirect public debate into disputes over methodology and circumvent external demands through internal agency or cross-agency coordination. Chidambaram's efforts to push the enumeration of caste into another state project (other than Census 2011) was the most visible sign of the behind-the-scenes negotiations underway. Methodological concerns became the central points of discussion even among caste census advocates.[61] Civil society organizations became relegated to a narrow role of commenting on methodology in policy formation or implementation. At the same time, the executive bureaucracy worked to create additional operational latitude for how to collect the caste-wise data, as I detail in the next section. The final step involves "expert closure," or the limiting of non-expert participation in subsequent deliberations over policy. During this stage, experts implement methodological and technical decisions, but in doing so violate the spirit of the original concession. The next two chapters describe the implementation of the caste census in another state project.

Reframing the caste census debate as technical intensified following the public concession. Home minister Chidambaram and the ORGI continued to oppose a full caste-wise enumeration in the census on technical grounds. The home minister argued in a cabinet meeting in late May 2010 that "the census must be protected" and asserted that enumerating caste would ruin the quality of census data. Chidambaram maintained that the decision should be apolitical and guided by the expertise of the ORGI. However, technical expertise within the state is hardly caste-free given the composition and organizational culture of the upper tiers of the bureaucracy, as discussed in Chapter 1. Chidambaram suggested that caste data instead be collected during the biometric capture phase of the National Population Register (NPR)—a new project to create a biometric national identity card. In doing so, he put forward the first plausible alternative to collecting caste-wise data in Census 2011. Key players in the executive bureaucracy also adopted Chidambaram's strategy to sidestep public ideological debates and front-stage technical concerns. In response to disagreement over how to move forward, the prime minister appointed a Group of Ministers (GoM) headed by finance minister Mukherjee in early June 2010 to examine "the enumeration of caste, other than Scheduled Castes and Scheduled Tribes, in Census 2011" and submit a report "expeditiously" to the prime minister.[62]

Trusted Political Leader Creates Operational Latitude and Narrows Dissent

For the next three months, finance minister Mukherjee played a key role in helping to revise the political consensus. He was head of the GoM and also the main point of contact with caste census advocates. He was a trusted leader who became the representative of the political leadership during negotiations with caste census advocates.

The GoM had three meetings between July and August 2010. The first meeting brought into focus the fact that there were divisions within the group on whether to include the caste-wise enumeration in Census 2011.[63]

The group decided to consult with other political parties before deciding how to proceed. The GoM presented this move as collaborative—that is, one that would provide additional confirmation of the cross-party consensus that appeared in the Lok Sabha. It also created political cover to ensure that no surprise dissenters appeared at a later stage. The finance minister wrote to every political party and asked the leadership of the party to respond in writing to two questions:

> (1) Whether "caste" should be canvassed in the ongoing Census/National Population Register (NPR) exercise? (2) If the answer to [the first question] is in the affirmative, whether your party agrees that the caste of the respondent should be canvassed in such a manner and at such a stage that it does not affect the integrity of the headcount (census)?[64]

The first question presented two legally and operationally distinct projects as interchangeable. The decennial census falls under the legal purview of the 1948 Census Act, while the NPR is governed by the Citizenship Rules of 2003. Data collection under the Census Act is confidential, and households are required to provide information. Census data provide the population total for the country and specific groups, which inform a range of policies and programs including the size of SC and ST reservation quotas. In addition, the decennial census is a widely respected government project with a 120-year history. The NPR, in contrast, is a first-time data collection effort to create a system of national identification. Caste census advocates demanded the enumeration of caste-wise data in the decennial census because of its history, stature, and ongoing role in determining reservation quotas. However, an affirmative response to the first question authorized the bureaucracy to collect caste data in the decennial census or in the NPR despite their legal, historical, and operational differences. The wording of the first question created operational latitude for how to collect caste-wise data, instead of simply securing an affirmative or negative response to the inclusion of a caste count in Census 2011.

The phrasing of the second question made it difficult for parties to answer "no." A political party was unlikely to state that it would like to see the "integrity" of the decennial census affected. Yet political parties also lacked technical expertise to challenge the premise of the question—that the collection of caste-wise data could somehow impact the integrity of the census. Assenting to the second question created additional leeway for the project, including the latitude to move the caste-wise enumeration out of the ORGI's jurisdiction altogether.

The wording of both questions shifted the focus from documenting political support for the collection of caste-wise data in Census 2011 to creating flexibility for how the bureaucracy would collect the data. The questions reveal the interests and influences of the home minister and the ORGI. Not surprisingly, every major political party, including the BJP, provided an affirmative response to both questions by the early August deadline.[65] Their assent authorized operational latitude for the project.

Congress Party president Sonia Gandhi, prime minister Manmohan Singh, and the cabinet began to forge a new path. The GoM met again in August 2010 and recommended that a full caste count occur during the biometric phase of the NPR, which was scheduled to start in November or December of 2010.[66] During this same period, finance minister Mukherjee's public announcements continued to describe the political leadership's plan as including a caste-wise enumeration in the census, which conflated two operationally and legally distinct projects.[67] The leadership used the language of "census" while intending to collect caste-wise data in the NPR. But that was not the end of the story; organized dissent by caste census advocates soon followed.

A subset of MPs spoke out vehemently against the GoM's recommendation. Yet, unlike the parliamentary debate in May 2010 that garnered widespread support across the political spectrum, this time the voices of OBC politicians dominated the opposition to inserting a caste count in the NPR. The narrowing of public dissent aligned with the written support that political parties had provided in which they assented to operational latitude to protect the "integrity" of the census. The dissenting MPs emphasized that the GoM's announcement no longer aligned with the demand of caste census advocates. On August 11, 2010, MPs Mulayam Singh Yadav and Lalu Prasad refused to let the late morning "Question Hour" begin in the Lok Sabha. Mulayam Singh shouted, "They are fooling us" and the "biometric stage will never come." Lalu Prasad argued that the biometric process will take "100 years and will still not be completed."[68] Sharad Yadav also joined the protests and urged house speaker Meira Kumar to allow them to discuss the matter and asked the Congress-led government to explain its position. Pressed into a corner, finance minister Mukherjee responded by suggesting that while the GoM had made its recommendation, the cabinet had yet to finalize its decision:

> The decision taken by the GoM is that caste will be collected in the census without affecting the integrity of the head count. How and when this should

> be done is under consideration. Some suggestions have been made by the leaders today. All these will be kept in view and an appropriate decision on the mechanism will be taken shortly. As the decision will have to be taken by the Cabinet, I will be able to inform the House only after the Cabinet takes the decision.[69]

Under competing pressure from OBC politicians and the ORGI, finance minister Mukherjee continued to waver back and forth in his statements in the months that followed. Several caste census advocates continued to view him as sympathetic to their demands, and later concluded that the interests of the ORGI and the power of the Home Ministry dominated and decided the outcome of internal bureaucratic negotiations. At the same time, public opposition had narrowed to political parties with OBC leaders unlike the unified outcry during the Lok Sabha debate in May 2010. The debate had shifted to seemingly technical discussions where census bureaucrats had much greater authority. The neutrality and objectivity of the decennial census became the rallying point, and "integrity" became framed with narrowly technical considerations instead of one with robust ethical and normative dimensions.[70]

Public intellectuals also expressed their dissent at the GoM's recommendation to include the caste-wise enumeration in the NPR and pushed the government to follow through on its May 2010 commitment.[71] Participants of a June 2010 conference on the caste census at the Centre for the Study of Social Exclusion and Inclusive Policy at the National Law School in Bengaluru, which was organized by professor and director S. Japhet and associate professor Chandan Gowda, published a dissenting open letter in *The Hindu* on August 13, 2010.[72] In the letter, the authors welcomed the GoM's "decision to enumerate caste in Census 2011" as a "progressive and much needed step towards re-orienting our polity and revitalising the implementation of social policy."[73] They objected, however, "to entrusting caste enumeration, not to the Census organisation but to biometric data capture."[74] They pressured the GoM "to reconsider this move because it will not only defeat the very purpose of enumerating caste but will condemn the entire exercise to almost certain failure."[75] The letter laid out several shortcomings to enumerating caste in the NPR, including its legal and operational distinction from the census. The debate now shifted to one fully focused on methodology, with advocates for a caste census asserting that modifications to the decennial census would not affect the integrity of census data and would produce reliable caste-wise data. Along with organizing conferences and writing op-eds, the public intellectuals in support of a caste-wise enumeration published the book *Caste Census: Towards an Inclusive India*, which contained many of the papers presented at the National Law School conference. At the book release in August 2010 in Delhi, the authors met with the prime minister and Congress Party leaders to push for their demands.[76]

On the final day of the monsoon session of parliament in late August 2010, politicians in support of a caste-wise enumeration in the census demanded a clear commitment from the Congress leadership. They pressed the finance minister to clarify Congress's position. MP Sharad Yadav pointed out that the government had sought the views of parties, the parties had responded in writing, and the finance minister had held three meetings with party leaders. Yet unease remained because the Congress leadership had not implemented a change in census policy on caste. MP Mulayam Singh similarly reminded Mukherjee that due to the consensus across political parties, the Congress leadership had promised to include a caste enumeration in the census but had not done so.

The finance minister once again assured the parliamentarians by saying that "the political parties had supported the idea of including caste in the census and there was no need for any 'apprehensions' over the issue."[77] Mukherjee explained that only the formalities remained as political consensus existed and the enumeration of caste would take place during the door-to-door enumeration of Census 2011, after a formal endorsement at the next cabinet meeting. The politicians and activists in support of a caste census were once again relieved by Mukherjee's clear statement of support.[78] They remained confident that a caste-wise enumeration would occur during the second round of the decennial census in early 2011.[79]

Float Plan Aligned with Demands to Avoid Criticism

The ORGI and the Home Ministry continued to work behind the scenes to keep the full caste-wise enumeration out of the decennial census. In early September 2010, home minister Chidambaram put forward a second alternative to enumerating caste in Census 2011:

> After considering various options, the option that we have approved is, based on the responses of various political parties, that caste must be canvassed, and the integrity of the headcount must not be affected.... A separate house-to-house enumeration of caste will be done during the period June 2011 to September 2011.

A standalone caste census aligned with the consensus to have the ORGI collect the caste-wise data while protecting the "integrity" of the decennial census.[80] By collecting the caste data in a separate exercise overseen by the ORGI, the agency would design and manage all field operations, and the central government would constitute an expert group to classify the caste data.[81] Yet a second census by the ORGI seemed financially and logistically impractical given that the massive decennial census operations would conclude just a few months prior to the start of the proposed standalone caste census. During the announcement, Chidambaram explained that the costs of the separate caste census would be assessed later. The home minister also remained evasive about whether data from the decennial census could be merged with the separate caste count. In this new alternative, the ORGI created a plan to collect caste-wise data that appeared to satisfy the interests of key parties, although it was financially and operationally impractical. Still for several months, until the government completed Census 2011, the public story remained that the ORGI would collect caste-wise data in a separate project following the decennial census. Yet, surreptitiously, the Congress-led government was developing a third (more practical) alternative that differed from the demands of caste census advocates.[82]

Punt Caste Count to Peripheral Location in Bureaucracy

The National Advisory Council, chaired by Congress Party president Sonia Gandhi, recommended combining the caste-wise enumeration with the BPL survey.[83] State governments in collaboration with the MoRD had previously conducted BPL surveys to identify rural households living below the poverty line with the goal of connecting eligible households to subsidized food and other anti-poverty programs. The National Advisory Council and Gandhi advocated that the ORGI could join forces with the MoRD to carry out the caste-wise enumeration as part of the BPL survey. They hoped that under the ORGI's jurisdiction many of the previous errors in the BPL survey, such as the exclusion of extremely poor households from BPL identification, would be reduced and the government could save resources by combining both projects. The MoRD had already piloted the BPL survey in rural areas, and the survey was scheduled to begin in 2011.[84]

Prime minister Singh sought the advice of the Economic Advisory Council to review the National Advisory Council's recommendation to combine both projects. The Economic Advisory Council disagreed with the recommendation. Chakravarthi Rangarajan, chair of the Economic Advisory Council, argued that conducting the BPL survey along with the caste-wise enumeration would delay the availability of caste-wise data and "attract interest groups which may impact the integrity of the census and caste enumeration."[85] The ORGI also had no interest in overseeing a joint BPL–caste census exercise. Publicly, an ORGI official argued, "[C]aste enumeration is individual based while the other two (socio-economic and economic census) are individual, household, and enterprise based. In view of this, caste enumeration should be done as a stand-alone exercise."[86] However, the ORGI neither wanted to design and implement a standalone census at the same time as Census 2011 nor be involved in the widely criticized BPL project. In the end, the executive leadership conceded to the ORGI's staunch opposition to oversee the joint caste census and BPL survey.

The government remained publicly silent on its shift away from a separate caste census until data collection for Census 2011 was complete in March 2011. Once the collection of caste-wise data in the census was no longer an option, the political leadership announced that they would include the caste-wise enumeration in a revised BPL survey that would be conducted in rural and urban areas between July and December 2011.[87] They renamed the BPL survey the Socio-Economic Caste (SEC) Census and designated the MoRD and the Ministry of Housing and Urban Poverty Alleviation (MoHUPA), the latter of which had not been involved in past BPL surveys, as the two nodal agencies for the project.[88] The ORGI would assist the project by providing maps and household listings for each enumeration block, overseeing the training of enumerators and field supervisors, and sharing household data from the NPR.

The ORGI and the Home Ministry succeeded in their effort to distance themselves from primary responsibility for the collection of caste-wise data. Their concern that a caste-wise enumeration would somehow damage the "integrity" of the decennial census proved an effective barrier to changing census policy on caste. The final plan pushed the caste-wise enumeration out of the ORGI's jurisdiction and under the authority of an agency without expertise and experience in coordinating and conducting a nationwide census. The MoRD would now take the lead and collect caste-wise data as part of a project that had been extensively critiqued in its earlier iterations.[89] This arrangement signaled the executive bureaucracy's low regard for the caste-wise data. The removal of the caste count from the jurisdiction of the well-resourced and technically skilled ORGI was consistent with the executive bureaucracy's minimal commitment to anti-caste programs and policies.

In independent India, most anti-caste programming has fallen under the jurisdiction of the Ministry of Social Justice and Empowerment, the

National Commissions for Scheduled Castes and Scheduled Tribes, and other ministries created to address discriminatory systems and indignities faced by marginalized groups. These efforts are concentrated and largely limited to organizational units that are explicitly related to caste and peripheral to the center of power, while the most powerful ministries promote castelessness. This pattern is consistent with the caste-elite composition of the upper tiers of the bureaucracy. Gopal Guru describes the displacement of Dalit bureaucrats to the Social Justice Ministry, Scheduled Caste Commissions, Directorate of Social Welfare, and similar state and local agencies that are "relatively less hostile" to the "cultural and mental well-being" of Dalits.[90] He explains how, "according to the hegemonic designs of the powerful," Dalit bureaucrats must "remain chained to those institutions that deal with Dalit issues" to "contain Dalit aspirations" and "insulate other institutions from the Dalit 'menace.'"[91] Executive bureaucrats continued a long history of compartmentalizing anti-casteist efforts to peripheral locations within the bureaucracy by punting the collection of caste-wise data to the BPL survey and the MoRD and the MoHUPA.

Looking Back and Ahead

In early May 2010, it appeared that for the first time in seventy years the central government would collect nationwide caste-wise data in an upcoming decennial census. Yet, in the months that followed, the internal bureaucratic machinery led a quiet rebellion that re-envisioned how to collect the caste-wise data. The Congress leadership secured written political support for the project from opposition parties, but in doing so also acquired operational latitude. Three different political maneuvers drove the caste count out of the decennial census and into the NPR, then into a standalone census exercise by the ORGI, and finally into a revamped BPL survey. Only after the ORGI completed Census 2011 did the political leadership publicly announce that the caste-wise data would be collected as part of an upcoming BPL survey. This trajectory moved the project out of the primary jurisdiction of the ORGI.

Advocates for a caste census had consistently demanded a caste-wise enumeration in the decennial census. Instead, the executive bureaucracy inserted the caste count in a project that had been highly criticized in its previous iterations and, in doing so, pursued a path likely to produce unreliable data. This was particularly ironic given that technical concerns over the "integrity" of the decennial census data were the primary justification for resisting a caste-wise enumeration in the census. Executive bureaucrats were not equally concerned about the "integrity" of the caste data. In sum, a grassroots mobilization secured a public commitment to change census policy on caste, only to have the policy change stalled in the less visible spaces of the executive bureaucracy.

Bureaucratic deflection involved six interconnected steps (see Table 3.1). Not every step was successful, and most emerged in response to organizing by caste census advocates. The strategy of shifting responsibility from the central government to the state level had been long-standing in relation to the collection of caste-wise data. When this strategy proved ineffective, the next several steps involved reframing decision-making as purely technical, which shifted the remaining decisions to experts. Executive bureaucrats moved the conversation away from the nature of caste-based power by elevating operational and technical considerations, which they presented as apolitical decisions. When it was no longer possible to collect caste-wise data in the census, the political leadership announced its plan to punt the caste count to the problem-riddled poverty survey and continued the long-standing practice of compartmentalizing anti-caste efforts to peripheral locations in the bureaucracy. The restriction of anti-casteist policies and programs to

Table 3.1 Institutional Strategies of Bureaucratic Deflection to Maintain Status Quo in Census Policy on Caste

Strategies that kept caste census out of Census 2011 and pushed it into another state project (i.e., revamped BPL survey)
Shifted responsibility from central government: decentralized decision-making to state governments or state-level backward classes commissions
Depoliticized the debate and focused on technical concerns: shifted debate from ideological differences to technical considerations and methodology, which were presented as apolitical decisions
Silenced non-experts: debate reframed as technical, so opinions of non-experts not required
Secured consent and operational latitude: while confirming consent created flexibility in how to carry out project
Bought additional time for planning (without external scrutiny): floated (impractical) plan aligned with original demands to buy time and avoid inspection/scrutiny
Compartmentalized anti-caste efforts to peripheral locations in bureaucracy: punted project to under-resourced, under-skilled, and/or peripheral unit

Source: Author.

under-resourced and less powerful locations in the bureaucracy emerged as a formidable strategy and supported a subjectivity of castelessness, which views caste as a problem of oppressed groups and obscures the power of privileged castes.

Bureaucratic deflection was also supported by allied institutions and groups. Celebrities and public intellectuals put forward the view that a caste-wise enumeration was anti-Indian and anti-national. Their slogan "my caste is Indian" was not new by any means. Anti-caste activists had more than a 100 years earlier—and across multiple locations—critiqued caste by organizing oppressed groups to answer in a particular way when asked about caste in colonial censuses. These organizers—usually from oppressed-caste backgrounds—fought to destroy caste hierarchy and dismantle Brahmanical ways of thinking. In contrast, the main coordinators of the *meri jati Hindustani* campaign promoted an uncritical nationalism in which they homogenized dissimilar types of caste consciousness. There was little room for distinctions between caste consciousness that was anti-casteist in its orientation versus that which reinforced caste hegemony. They framed efforts to produce a detailed caste-wise map of the distribution of socioeconomic power and resources as anti-Indian. In the process, they normalized casteist, Hindu nationalist ideologies, institutional norms, and practices as "Indian," and the efforts to change these norms as anti-nationalist and casteist.

The Supreme Court refused to intervene in petitions advocating for a change in census policy on caste. During one refusal to hear a petition supported by the lower courts, the Supreme Court portrayed a caste census as potentially dangerous—something that could create social strife. In this depiction, the change in census policy was depicted as possibly harmful, but the status quo was not critiqued, which neutralized pressure for change. Several state high courts saw the collection of caste data by the ORGI as necessary for implementing reservation policies, and the existing policy as deeply political, while the Supreme Court's relative autonomy allowed it to seemingly align with the idea put forward by the Home Ministry and the ORGI that the collection of caste-wise data in the census would politicize an otherwise apolitical project.

The relative silence of academia is also worth noting here. While a small group of scholars and public intellectuals spoke out forcefully in support of a caste-wise enumeration (and met with political leaders and joined the campaign) and others critiqued the project, most remained silent on the topic.[92] Perhaps this should be unsurprising given the caste composition of research scholars. Recent data gathered through right-to-information laws highlight the dearth of representation of oppressed castes among faculty at India's elite institutions—where comparatively fewer teaching responsibilities and greater research-related resources support research and scholarly writing and engagements. At the top-ranked Indian Institutes of Technology (IITs), fewer than 1 percent of professors come from SC and ST communities.[93] Data from the thirteen Indian Institutes of Management (IIMs) reveal that only 4 out of 642 faculty members are SCs, while only 1 out of 642 faculty members is ST. At the University of Delhi, only 3 out of 264 faculty members are SC.[94] Data on the pipeline of future faculty reiterates the persistence of the problem. With regard to postdoctoral funding, 80 percent of recipients of faculty fellowships from one of the main government science funding institutions were from privileged castes, while just 6 percent were SCs and less than 1 percent were STs between 2016 and 2020.[95] Similarly, among Ph.D. students at 13 elite research institutions, students from SC, ST, and OBC communities were well below their proportion in the population.[96] The Muslim community's enrollment as students in higher educational institutions has declined in recent years, and less than 5 percent of teachers in higher education are Muslim—a gross underrepresentation of the religious community.[97] These statistics trace the ongoing dominance of caste elites in higher education and the exclusion of oppressed castes and Muslims, particularly among faculty at research institutions. Entire disciplinary fields that could analyze caste-wise data and publish research that makes visible a system of domination and its ongoing effects remained largely silent.

Bureaucratic deflection and the support of allied institutions kept the caste-wise enumeration out of Census 2011. As the next two chapters document, a nationwide enumeration of caste took place between 2011 and 2014 as part of a revised BPL survey. This project was fraught with delays and coordination challenges, and the process of collecting caste data bore little resemblance to the 2010 demands of caste census advocates.

Notes

1. M. Vijayanunni, "Counting Caste: Methodology," in *Caste Census towards an Inclusive India*, ed. S. Japhet and Chandan Gowda (Bangalore: Centre for the Study of Social Exclusion and Inclusive Policy, National Law School of India University, 2010), 39–50.
2. Bureaucratic deflection refers to the process by which the administrative bureaucracy silently maneuvers, avoids, and resists the implementation of

policies and programs that challenge Brahmanical patriarchy and Hindutva ideologies, as defined in Chapter 1 and elaborated in Chapter 2.

3. Dilip Mandal, Interview by Gurinder Singh Azad, Anup Vimal, Noopur, and Anoop Kuman, *Dalit News from Kerala*, September 21, 2010, https://dalitskerala.wordpress.com/2010/09/21/, accessed April 1, 2024.
4. "Lok Sabha (No. 14) Debate" (New Delhi: Lok Sabha Secretariat, October 3, 2008), 570.
5. At the time, there were 545 MPs in the Lok Sabha and 245 in Rajya Sabha.
6. "Rajya Sabha Debate, Session 218" (New Delhi: Rajya Sabha Secretariat, December 9, 2009), 330.
7. "Rajya Sabha Debate, Session 218" (New Delhi: Rajya Sabha Secretariat, December 18, 2009), 263.
8. Rajindar Sachar et al., *Social, Economic and Educational Status of the Muslim Community of India: A Report*, Prime Minister's High Level Committee (New Delhi: Cabinet Secretariat, November 2006), https://www.education.gov.in/sites/upload_files/mhrd/files/sachar_comm.pdf, accessed April 1, 2024.
9. Ajay Maken, "Document No. H-1102/8/2010-CD(CEN)" (Ministry of State for Home Affairs, February 20, 2010).
10. The political leadership of the BJP used identical language in relation to Census 2021, as described in the closing chapters of this book.
11. Interview with activist, May 9, 2013.
12. "Ramadoss Seeks Caste-Wise Census," *The Hindu*, September 19, 2009.
13. Ibid.; J. Venkatesan, "Supreme Court Declines to Entertain PMK's Petition," *The Hindu*, April 10, 2009.
14. The PMK's counsel responded that several representations had been filed but no decision had been made by the government—hence the legal petition. The Chief Justice replied with humor, "Wait for the new government. You may be in that." The Supreme Court refused to mandate a change in census policy on caste by arguing that policymaking was outside its jurisdiction. Venkatesan, "Supreme Court Declines to Entertain PMK's Petition."
15. "Plea for Caste-Wise OBC Enumeration," *The Hindu*, October 16, 2009; "Issue of Caste-Wise Census Taken Up with Centre," *The Hindu*, January 24, 2010.
16. Throughout much of south India, some form of affirmative action had existed since the 1920s, and state courts supported the demand to collect caste-wise data to determine the size of reservation quotas and aid in the implementation of reservation policies.
17. J. Venkatesan, "Apex Court Moved for OBC Census," *The Hindu*, April 15, 2010.
18. "Court Directs Government to Conduct Caste-Wise Census," *The Hindu*, May 13, 2010.
19. Interview with activist, June 27, 2015.

20. Ibid.
21. Vijay Kumar, "Demand for Caste-Wise Census in 2011," *The Hindu*, December 24, 2009.
22. Interview with activist, July 17, 2016
23. Interview with activist, June 22, 2015.
24. Interview with activist, July 17, 2016.
25. Jaffrelot and Kumar, *Rise of the Plebeians?*; Jaffrelot, *India's Silent Revolution*; Yadav, "Reconfiguration in Indian Politics"; Yadav, "Understanding the Second Democratic Upsurge."
26. Interview with politician, June 24, 2015.
27. Ibid.
28. The household listing phase of the decennial census involves enumerators visiting every household in the country to create a neighborhood/village level block map and a list of every household in the block/village to create baseline data for the official census count in early 2011.
29. "Lok Sabha (No. 15) Debate" (New Delhi: Lok Sabha Secretariat, May 6, 2010), 413–14.
30. Ibid., 407.
31. Ibid., 388.
32. Ibid., 397.
33. Neena Vyas, "No Harm in Caste-Based Census: BJP," *The Hindu*, May 5, 2010.
34. "Lok Sabha (No. 15) Debate" (New Delhi: Lok Sabha Secretariat, May 7, 2010), 557.
35. Some survey experts did speak out in support of a caste census, but they were not consulted to implement a caste-wise enumeration. See experts who contributed to the volume: S. Japhet and Chandan Gowda, eds., *Caste Census towards an Inclusive India* (Bangalore: Centre for the Study of Social Exclusion and Inclusive Policy, National Law School of India University, 2010).
36. The Congress leadership at the time criticized prime minister V. P. Singh's implementation of the Mandal Commission report. Rajiv Gandhi, the then president of Congress, accused the prime minister of bringing "the country to the edge of caste war" for his expansion of caste-based reservations to OBCs; Gandhi instead put forth alternative economic criteria arguing that "assistance should be given to the truly poor, to the landless, to the people falling in the poorest category." *Indian Express*, September 8, 1990, as cited in Jaffrelot, *India's Silent Revolution*, 428.
37. Jaffrelot, *India's Silent Revolution*, 427; Paul Brass, "The Politicization of the Peasantry in a North Indian State," *The Journal of Peasant Studies* 7, no. 4 (n.d.): 395–426.
38. Congress regained power in 1991 and prime minister Narasimha Rao issued an executive order—revising former prime minister V. P. Singh's

executive order that extended reservations to OBCs—that instead reserved 10 percent of posts in government jobs for economically weaker sections of groups not already covered by quotas—thereby excluding SCs and STs. However, Congress adjusted its position after judges decided in 1992 that solely economic criteria could not be used to define "backwardness." Jaffrelot, *India's Silent Revolution.*

39. "Lok Sabha (No. 15) Debate" (New Delhi: Lok Sabha Secretariat, May 7, 2010), 570.
40. Sunil Prabhu, "Pranab Says Current Census Will Include Caste, after All," NDTV (May 7, 2010); "Caste Census Should Have Continued after Independence: Pranab," *Times of India*, May 8, 2010.
41. "Census 2011 to Include Caste," *The Hindu*, May 8, 2010.
42. Bachchan is a household name throughout India due to his illustrious movie career and never-ending presence on television ads and billboards across the country. Bachchan's interview in May 2010 was part of the household-listing phase of Census 2011, with the main census round to occur in early 2011. Amitabh Bachchan, "Day 750," *Bachchan Bol* (blog), May 11, 2010, http://srbachchan.tumblr.com/post/25143096756, accessed April 1, 2024.
43. Ibid.
44. While Bachchan was correct that his opinion about caste in the census would be of widespread interest, he was incorrect in assuming that the shift in census policy would affect the household-listing phase of the decennial census that was underway in May 2010. Instead, the much-debated changes were anticipated to take place in the main round of Census 2011 in February/March 2011.
45. Ibid.
46. Barkha Dutt, "In Reverse Gear," *The Hindustan Times*, May 14, 2010.
47. Pratap Bhanu Mehta, "My Caste and I," *The Indian Express*, May 12, 2010.
48. Nehru spoke against the caste system, which was "opposed to democratic conception" and kept people "to their hereditary stations and did not challenge the social order." Yet Nehru also believed that despite the caste system's flexibility and longevity it would be upended by "basic economic changes which have shaken up the whole fabric of Indian society"—that is, caste would diminish in importance against the background of economic growth and development. Nehru focused on promoting a unifying Indian identity instead of dismantling caste hierarchy. Jawaharlal Nehru, *The Discovery of India* (New York: Anchor Books, 1960), 148, 162.
49. P. Radhakrishnan, "Caste Factor in Census 2011 Will Pollute India's Governance," *DNA*, July 12, 2011; Mehta, "My Caste and I."
50. HT Correspondent, "Write Meri Jaati Hindustani in 2011 Caste Based Census"; HT Correspondent, "Civil Society Movement Opposes Caste Census."

51. Other prominent individuals connected with the campaign include former cabinet minister Arif Mohammad Khan, former ambassador J. C. Sharma, activist Alka Madhok, and Supreme Court lawyers J. C. Batra and former law minister and MP Ram Jethmalani.
52. Radhakrishnan, "Caste Factor in Census 2011 Will Pollute India's Governance"; Staff Reporter, "Movement against Caste-Based Census Launched"; *The Hindu*, "Movement against Inclusion of Caste in Census Gains Impetus."
53. Gopal Guru, "The Indian Nation in Its Egalitarian Conception," in *Dalit Studies*, ed. Ramnarayan Rawat and S. Satyanarayana, 31–51 (Durham, NC: Duke University Press, 2016), 34.
54. Ajay Maken, "Document No. H-1102/8/2010-CD(CEN)" (Ministry of State for Home Affairs, February 20, 2010).
55. Ibid.
56. Sujay Mehdudia and Siddharth Varadarajan, "Government Not for Caste Census," *The Hindu*, May 5, 2010, https://www.thehindu.com/news/national/Government-not-for-caste-census/article16298373.ece, accessed April 1, 2024.
57. Law minister Veerappa Moily, urban development minister Jaipal Reddy, minister for overseas affairs Vayalar Ravi, and telecom minister A. Raja supported a caste-wise enumeration in Census 2011. Law minister Veerappa Moily spoke out for the need to resume the caste census that occurred during the colonial period and publicly asked prime minister Singh to consider this change in census policy. Probably due to push back from the home minister, prime minister, and Congress president, Moily's opposition became less visible and finally completely silent to the broader public as time progressed. J. Venkatesan, "Moily Defends Caste-Based Census," *The Hindu*, May 21, 2010, https://www.thehindu.com/news/national/Moily-defends-caste-based-census/article16302260.ece, accessed April 1, 2024.
58. Mehdudia and Varadarajan, "Government Not for Caste Census."
59. "Lok Sabha (No. 15) Debate" (New Delhi: Lok Sabha Secretariat, May 7, 2010).
60. Trina Vithayathil, Diana Graizbord, and Cedric de Leon, "The Retreat to Method: The Aftermath of Elite Concession to Civil Society in India and Mexico," *Studies in Comparative International Development* 54, no. 1 (2019): 19–39.
61. Among those supporting an expanded caste count, most favored a count of all castes while others argued that the decennial census should simply add OBCs to the existing practice of enumerating SCs and STs. Yogendra Yadav, a scholar of Indian elections, argued for this incremental step in his push for a policy change. He asserted that the addition of the OBC category to the question on caste in Census 2011 was a straightforward insertion that would align the instrument with existing state policies. Other caste

census advocates argued that a complete social map was needed to inform the development and administration of programs and policies to reduce caste-based social and economic inequalities and promote a broader anti-casteist agenda—and therefore a full enumeration of all subcastes should occur in Census 2011. Yogendra Yadav, "Why Caste Should Be Counted," *The Hindu*, May 14, 2010.

62. "Mukherjee to Head GoM Caste-Based Census Panel," *Deccan Herald*, June 9, 2010.
63. In the first meeting on July 1, 2010, among the seven ministers who were present, finance minister Mukherjee, social justice minister Wasnik, and minister of state for minority affairs Khurshid spoke in favor of the caste census. Wasnik pointed out that prime minister Singh had already made a commitment on the floor of parliament to include a caste enumeration in the decennial census. In contrast, defense minister A. K. Anthony, home minister Chidambaram, and human resources development minister Kapil Sibal spoke out against a caste-based census. One of the ministers, who had previously voiced strong support for the caste census, law minister Moily, had another appointment midway through the meeting and three cabinet members did not attend (agriculture ninister Pawar, energy minister Abdullah, and railway minister Banerjee). "GoM Divided over Caste Census," *The Times of India*, July 1, 2010.
64. Neena Vyas, "BJP Ducks Firm Response to Pranab Query on Census," *The Hindu*, August 7, 2010.
65. During the month of July, while the GoM waited for responses from the political parties, finance minister Mukherjee met with the leaders of coalition and opposition parties. The Samajwadi Party, Janata Dal-United, and Rashtriya Janata Dal remained in support of the caste census. In contrast, the BJP's position seemed to waiver back and forth. On July 2, a top BJP leader categorically said the party would reiterate its support for a caste count, but later in the month BJP leader Nitin Gadkari publicly spoke out against a caste census. Smita Gupta, "Deadline Nears for Parties to Take Stand on Caste Census," *The Hindu*, August 2, 2010; Smita Gupta, "GoM Approval for Caste-Based Census," *The Hindu*, August 11, 2010; Maneesh Chibber, "GoM Unanimous Yes to Caste in Census," *The Indian Express*, August 12, 2010; Aarti Dhar, "Caste in Census: Cabinet to Decide on Modalities," *The Hindu*, August 12, 2010; "Call on Caste Census by Cabinet," *Hindustan Times*, August 19, 2010.
66. Ibid.
67. Ibid.
68. Ibid.
69. "India Approves Caste Based Census," BBC News, August 12, 2010. Gupta, "GoM Approval for Caste-Based Census."

70. Interview with activist, February 22, 2013. Interview with activist, May 9, 2013. Interrogating concepts of integrity, objectivity, and expertise as they relate to science, policy, politics, and ethics has a long history with science and technology studies scholars. See Sheila Jasanoff, "Science, Politics, and the Renegotiation of Expertise at EPA," *Osiris* 7 (1992): 194–217; Sheila Jasanoff, *The Fifth Branch* (Cambridge, MA: Harvard University Press, 1993); Sheila Jasanoff, *States of Knowledge: The Co-Production of Science and the Social Order* (London and New York: Routledge, 2004).
71. Japhet and Gowda, *Caste Census towards an Inclusive India.*
72. M. Vijayanunni et al., "Letter to the Group of Ministers on Caste Census," *The Hindu*, August 13, 2010.
73. Ibid.
74. Ibid.
75. Ibid.
76. Interview with activist, June 17, 2015.
77. Smita Gupta and Aarti Dhar, "Stand-Alone Caste Headcount Too a Certainty," *The Hindu*, August 31, 2010; Aarti Dhar, "Now, Caste Census Is Just a Formality, Says Centre," *The Hindu*, September 1, 2010.
78. Interview with activist, July 17, 2016.
79. Interview with activist, June 22, 2015.
80. As cited in Smita Gupta, "Centre Clears Caste in Census 2010," *The Hindu*, September 9, 2010.
81. Press Information Bureau, Government of India (Cabinet), "Enumeration of Castes Other than Scheduled Castes and Scheduled Tribes," *News Release*, September 9, 2010, https://pib.gov.in/newsite/PrintRelease.aspx?relid=65631, accessed April 1, 2024.
82. K. P. M. Basheer, "Nature of Caste Census Still Unclear," *The Hindu*, December 13, 2010.
83. The National Advisory Council consisted of senior Congress leaders, retired bureaucrats, economists, and activists. Prime minister Singh created the National Advisory Council in 2004 and reinstated it in 2010 to advise him on policy matters and draft legislation. "Sonia as NAC Head Is Pseudo-Constitutional Power Center: BJP," *The Indian Express*, March 30, 2010.
84. Gargi Parsai, "It's for States to Identify BPL Families," *The Hindu*, January 14, 2011.
85. Ibid.
86. Ibid.
87. Press Information Bureau, Government of India (Ministry of Rural Development), "Below Poverty Line (BPL) Census to Be Conducted along with Caste Census during June–December 2011," *News Release*, May 19, 2011, https://pib.gov.in/newsite/PrintRelease.aspx?relid=72217, accessed April 1, 2024.

88. Ibid.
89. See F. Ram, S. K. Mohanty, and Usha Ram, "Understanding the Distribution of BPL Cards," *Economic and Political Weekly* 44, no. 7 (2009); Indira Hirway, "Identification of BPL Households for Poverty Alleviation Programmes," *Economic and Political Weekly* 38, no. 45 (2003): 4803–08.
90. Guru, "Liberal Democracy in India and the Dalit Critique," 111.
91. Ibid., 108.
92. Japhet and Gowda, *Caste Census towards an Inclusive India*; Vijayanunni et al., "Letter to the Group of Ministers on Caste Census."
93. Ankur Paliwal, "How India's Caste System Limits Diversity in Science—in Six Charts," *Nature* 613, no. 7943 (2023): 230–34.
94. Abhishek Hari, "Casteism Is Rampant in Higher Education Institutions, but Is 'Wilfully Neglected': Study," *The Wire* 8 (2021).
95. Paliwal, "How India's Caste System Limits Diversity in Science—in Six Charts," 233.
96. Ibid.
97. Vikas Pathak, "Only 4.9% of Higher Education Teachers Are Muslims," *The Hindu*, January 12, 2018; Ziya Us Salam, "Muslims Lag Behind Scheduled Castes and Scheduled Tribes in Higher Education," *The Hindu*, May 31, 2023.

4

Survey Making in an Era of Castelessness

> It is idle and futile to expect that a BPL survey will, just by giving it the bombastic misnomer "Socio-Economic and Caste Census 2011," become one. Census is an exclusively central subject (entry 69 in the Union list under the seventh schedule of the Constitution). It is only the Centre that can, by notification in the official gazette under Section 3 of the Census Act, authorise a census, and, without such a statutory backing, this poverty survey (rightly called a BPL survey) conducted by the state governments cannot be passed off as a census. This exercise has no socio-economic data coming out of it except bare poverty statistics, and, with just a question asking for caste inserted into it as a fifth wheel, it does not become a "caste census" by any standards, nor does it generate a caste-wise socio-economic profile of the population of India as required by the Supreme Court in the caste-reservation case ... this exercise defeats the whole purpose of a caste census.[1]
>
> —M. Vijayanunni, 2011

M. Vijayanunni, the former census commissioner, continued to speak out publicly against the combined caste-wise enumeration and BPL survey just before the project began in late July 2011. His prescient comments outlined the legal distinction between the decennial census and the BPL survey, related operational differences, and the shortcomings of the revised survey instrument. Vijayanunni pointed out that the SEC survey remained a project designed to identify those households living below the poverty line and differentiate among households that faced different degrees of food insecurity, housing insecurity, and poverty. While the project sought to pinpoint households above the poverty line, it did not seek to distinguish within this broad grouping. Detailing the socioeconomic profiles of comparatively privileged households fell outside the scope of the BPL identification. As such, degrees of privilege remained comparatively masked by the survey instrument. The survey did not seek to generate "a caste-wise socio-economic profile of the population of India" as required for the implementation of reservation benefits and related legal cases. Vijayanunni argued that inserting a caste question into a survey whose purpose was to identify poor households could not generate a comprehensive caste-wise socioeconomic profile of the country, nor a more detailed understanding of caste-based power. He continued to assert that only the decennial census had the legal authority to include a "caste census."

The BPL survey experienced an overhaul prior to the combining of the two projects. Facing pressure from the national *Right to Food* campaign that sought to universalize food access, reduce food prices, and expand household eligibility for subsidized food, the MoRD had been working to improve the BPL survey for several years. In 2008, Pradeep Jain, the minister of rural development, appointed an expert committee to develop "a more suitable methodology to identify the poor in rural areas" and address problems related to earlier rounds of the BPL survey, which included failing to enumerate and classify many of the poorest-of-the-poor households.[2] The expert committee, chaired by N. C. Saxena, a former IAS officer and past secretary for the Planning Commission and for the MoRD, had studied the shortcomings of the three previous BPL surveys and submitted its final report to the MoRD in August 2009. The document—widely known as the "Saxena Committee Report"— specified problems with past BPL surveys and laid out a revised methodology for identifying poor households.

This chapter traces key decisions made by experts involved in the planning and execution of the BPL survey as they first revised the survey to incorporate the recommendations of the Saxena Committee and then inserted a caste-wise enumeration into the BPL survey. I focus on three inter-related decisions made during the first half of 2011 that heavily shaped the on-the-ground process of collecting caste-wise data: revisions to the caste questions in the BPL questionnaire; shifting from the use of paper surveys to tablets during household interviews for "real-time" electronic data entry; and the related choice to outsource data entry to the private sector. I also discuss the decision not to have a centralized publicity campaign for the survey. In doing so, I show the widening gap between the demands of caste census advocates and the design and implementation of the caste-wise enumeration, as *bureaucratic deflection* continued to unfold.[3] The chapter also exposes how castelessness became embedded in the final survey design and related training materials.[4]

Revising the BPL Survey

A national debate and movement critiquing the government's role in addressing poverty, food security, and underdevelopment created a vibrant backdrop to the process of redesigning the BPL survey. Should access to food be a legal entitlement? Was it the government's responsibility to ensure basic sustenance of all people? The *Right to Food* campaign sought to universalize food access, reduce food prices, and expand household eligibility for the government's subsidized food distribution program. The campaign argued that access to affordable food—a key benefit of BPL status for households—should be legally entitled for all and pressured the government to pass a national food security bill.

Addressing the Saxena Committee's Recommendation

Against the backdrop of this high-profile campaign, the Saxena Committee—tasked with revising the methodology for conducting the BPL survey—documented its position on food access, shortcomings of previous BPL surveys, and recommendations for the upcoming BPL survey.[5] The committee proposed the inclusion of simple indicators to capture socioeconomic marginalization by caste, gender, class, disability, and other stigmatized identities and the use of these indicators to identify households in need of food subsidies and other poverty reduction programs. Their report states:

> The experience with earlier surveys [is] that the most poor are ironically often excluded from BPL survey lists, because of their social, economic and political powerlessness. What is even more distressing is the fact that a large number of them are not in any list—not even APL [Above Poverty Line]—and these must be the voiceless people living in remote and isolated hamlets, where roads do not exist, and for various reasons administration has decided to ignore them. Their exclusion goes beyond stark numbers and there is enough evidence to show that it is intrinsically linked to caste, gender and social inequality, and ownership of assets alike. People may be barred from access to food and livelihoods even if it is locally available and they have the economic means. They face insurmountable social barriers to food and livelihood security which may include gender, caste, age, ethnicity, disability or stigmatised ailments. Herein lie the foundations for the several social categories in our criteria for necessary inclusion. Their economic deprivation further derives from lack of productive assets like land and water, and access to credit.[6]

In their proposed methodology, the committee developed eight criteria to identify households for automatic inclusion on the BPL list.[7] The committee fully understood that local hierarchies and politics shaped the state's identification of poor households and therefore sought "to clearly identify the poor, through directly verifiable and observable characteristics (to make for administrative ease and convenience)—so they are not amenable to easy manipulation by local elite."[8] The creation of eight categories for automatic inclusion contrasted with previous BPL identification measures based on income and consumption. In addition, the automatic exclusion of households would continue as had occurred in past rounds of the survey, although the criteria had changed slightly, and the government would use the collected data to rank households that were neither automatically included in, nor automatically excluded from, the BPL list. Depending on available resources at the state level, a portion of the ranked households would also be eligible for BPL programs and related services.

The BPL administrative team within the MoRD headquarters in Delhi drafted a survey instrument based on the Saxena Committee's recommendations. In August and September 2010, the MoRD field-tested the pilot BPL survey in a two-step process across a sample of 254 villages.[9] In each village, a field supervisor first completed a village-level survey and then interviewers visited every household to complete a household survey. Both the village and household surveys included questions on caste and religion, and the wording of questions and answer options signaled how BPL administrators conceptualized and operationalized caste in the pilot survey based on input from the Saxena Committee report.

First, BPL administrators envisioned caste both as a localized identity and as a government administrative category. The latter might be consistent across central and state government lists or might differ. In the sixth section of the village survey, the field supervisor was expected to manually "record the religion and name of caste/tribe of every group residing in the village using the help of key informants."[10] For each listed caste/tribe, the field supervisor then identified and recorded the government administrative category. The supervisor manual instructed the field supervisor to enter the associated central government caste/tribe subgroup in column 3 and the related state government caste/tribe subgroup in column 4 of the survey. The four central government caste/tribe category options were Scheduled Caste (SC), Scheduled Tribe (ST), Other Backward Class (OBC), and "General (others)."[11] The state government categories consisted of SC, ST, OBC, and additional choices that varied by state.[12] As such, BPL administrators operationalized caste as a localized identity, a state-level administrative category, and an all-India administrative category.

This operationalization of caste differed from past BPL surveys. Previous rounds had restricted data collection to SCs and STs. The redesigned BPL survey that the MoRD field-tested in 2010 began to dismantle the ideology of castelessness embedded in previous rounds of the survey by collecting caste-wise data from all rural households including households within the "general" category, as well as OBCs.[13] The revised village survey, and linked household survey, also addressed the Saxena Committee's recommendation to differentiate among groups most stigmatized within the SC and ST categories. The pilot survey collected information on additional state-level categories such as Mahadalits (that is, most disadvantaged Dalit communities).

Second, BPL administrators continued the past practice of conceptualizing caste and religion as household identities. The piloted household survey consisted of a nine-page questionnaire that collected extensive household data and limited individual-level data. The first section entitled "Household Basic Characteristics" included three questions that asked for the name of the head of household, caste/tribe, and religion (see Figure 4.1). The question on caste/tribe was open-ended, with the expectation that the interviewer would record the *jati*, or the local endogamous kinship group associated with the household.[14] The interviewer manual states, "enter the full name of the caste/ tribe of the household in words. Do not write SC/ST/OBC/General, etc."[15] The instructions indicate that BPL administrators conceptualized caste as a household identity and the survey did not collect individual-level caste data.

<table>
<tr><th colspan="6">Block-1. Household Basic Characteristics</th><th>Central list (A)</th><th>State (B)</th></tr>
<tr><td>1</td><td>Name of Household Head</td><td></td><td>3</td><td>Caste/ tribe group</td><td>(post-code)</td><td></td><td></td></tr>
<tr><td>2</td><td>Name of Caste/Tribe</td><td></td><td>4</td><td>Religion</td><td>(code)</td><td></td><td></td></tr>
<tr><td colspan="8">Block-1 Codes</td></tr>
<tr><td colspan="8">Q3: Caste/tribe Group – POST-CODE (A) from Col.3; (B) from Col.4 of Block 6 in Village Survey</td></tr>
<tr><td colspan="8">Q4: Religion: 1=Hinduism; 2=Islam; 3=Christianity; 4=Sikhism; 5=Jainism; 6=Buddhism; 97=Other (specify)</td></tr>
</table>

Figure 4.1 Pilot BPL Survey: Caste and Religion in Household Questionnaire (2010)

Source: Ministry of Rural Development, "Socio-Economic Survey 2010: Supervisor Manual," 20.

The placement of the caste and religion questions within the "household section" of the questionnaire, as well as the investigator manual, conveys how the designers of the BPL survey viewed caste (and religion) as household identities. In contrast, the BPL administrators envisioned sex, age, and marital status as individual-level identities and included questions on these topics in a subsequent section of the survey entitled "Household Members Details."

Third, the structure of the pilot BPL survey also created a situation where the head of household's caste and religious identities became equated with the household's caste and religious identities. The investigator manual states: "if [there are] more than one caste/tribe in the same household (e.g., in case of inter caste marriages), report the caste/tribe of the household that the respondent identifies the household with."[16] Similarly, for the question on religion, the interviewer is instructed to "enter code as per list. In case of inter-religion marriages, write the code of religion as reported by the respondent."[17] However, the questions on religion and caste immediately followed a question about the head of household's name; the investigator would have no knowledge of whether other members of the household had a different religious or caste identity. Only later in the questionnaire are there questions about individual members of the household, but there are no questions about caste or religion. The interviewer would not know whether individuals with different caste or religious identities resided in the same household due to the design of the questionnaire. Questions about religion and caste immediately followed the name of the head of household, such that respondents were likely to share the religious and caste identity of the head of household. In addition, despite instructions to not presume the head of household was an adult male, widespread patriarchal social norms were likely to lead to the practice of the head of household being reported as an adult male (when present), and then their caste and religious identities would become the default for the household.

Finally, the designers of the BPL questionnaire envisioned that every household had a caste identity. The choice of "no caste" was not an imagined option. The investigator manual specifically considers inter-caste marriages and states that if "the household has no preferences, use the caste/tribe of household head."The BPL administrators instructed investigators to record the caste identity of all households, while the instructions slightly differed for religion.[18] For inter-religious households, the instructions state that "if no preferences are reported by the respondent, write the code of the religion of household head."[19] However, administrators did imagine the equivalent of a "no-religion" option as indicated by the instructions "in case of atheist or households not declaring the religion, please write 97 and specify."[20] In cases where the head of household did not believe in religion, such answers could be recorded as the household's religious identity. Households could "opt out" of providing data on religion, while the option to "opt out" of caste did not exist.

As such, the field-tested BPL survey conceptualized both religion and caste as household-level identities; the caste and religion of the head of household—usually an adult male—became the de facto identity for the household; and while not all households needed to provide information on religion, the survey designers expected that all households would provide data on caste. Aligning with the Saxena Committee's recommendations, the field-tested BPL survey challenged past practices of castelessness by enumerating localized caste identities for all rural households and classifying each *jati* by state- and central-government administrative categories. The design of the pilot BPL survey aided in processes of commensurating caste through the completion of the village survey, which required developing a local list of castes that standardized caste names within each village and classified them by state and central government administrative categories allowing for comparisons within a state and across regions.

In May 2011, the BPL administrative team found itself having to make additional changes to the already field-tested and revised BPL instrument after it was announced that a "caste census" would be inserted into the BPL survey. The survey instrument needed to be modified to collect individual-level information on caste, instead of household-level data. They also had to expand the data collection and processing plans and coordinate with two additional agencies—the MoHUPA and the ORGI. Both units had been appointed by the political leadership to collaborate with the MoRD in a combined BPL survey and caste-wise enumeration to take place in rural and urban India beginning in July 2011.

Incorporating the Caste Survey into the Revised BPL Survey

When the political leadership announced that they were combining both projects in May 2011, the BPL survey administrators had already redesigned and field-tested a version of the survey that incorporated the Saxena Committee's recommendations.[21] Now, a new survey instrument needed to collect individual-level data on caste and religion and expand its reach to include urban areas. While the piloted village and household BPL surveys could have been the basis for the individual-level questions about caste in the SEC survey, the Census of 2011 instead became the model. The decennial census had collected individual-level data on caste/tribe, but only for SCs and STs and not for OBCs and caste elites like the piloted BPL survey. The recently concluded decennial census neither created a caste list for the "general" category nor used existing OBC lists. The SEC survey designers now replaced the caste-related questions that emerged from the Saxena Committee's recommendations with questions that the ORGI recommended.

The finalized SEC survey included three individual-level questions on caste and religion, consisting of a question on religion with no pre-set categories and two caste-related questions (see Table 4.1). The first caste

Table 4.1 Religion and Caste Questions in the Pilot BPL Survey (2010), Census (2011), and SEC Survey (2011)

	Pilot BPL Survey	**Census 2011**	**SEC Survey 2011**
Religion	*Household level* Q4: Religion (code): 1= Hinduism; 2=Islam; 3=Christianity; 4=Sikhism; 5=Jainism; 6=Buddhism; 97=other and then specify name of religion.	*Individual level* Q7. Religion (write the name of the religion in full) Also give code in box if found in the list below. For other religions, write name of the religion in full but do not give any code number. 1… Hindu; 2…Muslim; 3…Christian; 4…Sikh; 5…Buddhist; 6…Jain	*Individual level* Q12: *Religion* (Write name of religion in full)
Caste	*Household level* Q2: Name of Caste/ Tribe: Q3: Caste/tribe group (post-code): Central list (A): State list (B): *Field supervisor post-codes caste group based on its classification in village survey, for central government list (SC, ST, OBC, general) and state list*	*Individual level* Q8. Scheduled Caste (SC)/ Scheduled Tribe (ST) *8(a) Is this person SC/ST?'* *If 'YES' give code in box* *SC….1* *ST….2* *If 'NO' put '3' in box* *8(b) If SC or ST write name of the SC or ST from the list supplied*	*Individual level* Q13: Caste/ Tribe Status Give code: Scheduled Caste (SC)-1; Scheduled Tribe (ST)-2; Other-3; No Caste/Tribe-4 Q_14: If code 1, 2 or 3 in Column 13, write names of caste/tribe. If code 4 in column 13, put 'X'

Source: Ministry of Rural Development, "Socio-Economic Survey 2010: Supervisor Manual"; Office of the Registrar General and Census Commissioner India, "Census of India 2011: Household Schedule," (Government of India, 2011); Ministry of Rural Development, "Socio Economic & Caste Census 2011 in Rural India" (Government of India, July 25, 2011), 31.

question had four answer options: SC, ST, other, and no caste/tribe.[22] Important changes to the answer options shifted the question further away from the piloted BPL survey. First, the answer options did not include the state administrative categories of "OBC" and "general," but added a "no caste/tribe" option. The second caste-related question sought to record subcaste information for all individuals except for those who responded "no caste" to the first question.[23] The training manual instructed data collectors to record whatever answer the respondent provided with no attempt to commensurate the data locally as in the pilot BPL—except for SC and ST classification. Table 4.1 compares the caste and religion questions across the census, pilot BPL survey, and SEC survey.

The revisions to the SEC survey ignored the original demands by caste census advocates and reversed several changes made to the caste-related questions in the BPL survey based on recommendations by the Saxena Committee and the national *Right to Food* campaign. At the same time, the primary purpose of the revamped BPL survey remained to identify households eligible for BPL status. The determination of a household's poverty status was based on household-level data.[24] As such, the instructions for questions that informed the automatic inclusion or exclusion of households from BPL status were detailed and specific, the questions were worded clearly, and the instructions and questions had been field-tested. The final caste-related questions in the SEC survey did not undergo a thorough field-testing, reinforced castelessness in the survey design, and had insufficient instructions regarding the caste-wise enumeration of religious minorities.

Castelessness in the Survey Design

The first question on caste made little effort to classify approximately 77.5 percent of the population into a state administrative category. The decision not to categorize individuals within the OBC and "general" administrative categories at the time of enumeration meant that "other" became a catch-all classification. Groups as diverse as Dalit Muslims and caste-elite Brahmans would fall into the default "other" category. In addition, at best, the "other" category would include thousands of castes that would require considerable time and effort to post-code into "OBC" and "general" categories, usually by individuals who had far less knowledge of local caste hierarchies. Or worse, without efforts to standardize the spelling of caste names and use pre-created caste lists for the second caste question, there could easily be millions of responses that would prove extremely challenging to post-code and classify.

Like with recent decennial censuses, the decision not to include all four major central government administrative codes in the first caste-related question, and instead only classify SCs and STs, reinforced an ideology of castelessness in the survey instrument.

The inclusion of a "no caste/tribe" answer option had a similar effect. It was not an option in the field-tested BPL survey nor in recent government surveys, so who did planners imagine would select this option and why was it added? The enumerator manual simply states, "[I]n case a person says that she/he has no Caste/Tribe, please record as 'no Caste/Tribe.'"[25] Later when providing an explanation for each answer option for the first caste question, the manual states, "[S]ome respondents may state that she/he or any member of her/his household has no Caste/Tribe. In such cases record Code - '4' under this question."[26] The enumerator manual envisioned that the "no caste" answer option would emerge from respondents after enumerators asked them about their "caste/tribe status." The types of households that might provide this answer remained unclear from the written instructions.

Two reasonable conclusions could be drawn from the broader purpose of the survey and the history and debates surrounding the enumeration of caste. First, the answer option of "no caste" was unlikely to apply to, or be offered up by, SCs and STs in this instance. SC/ST status is one of the seven "deprivation criteria" by which the government ranks households that are not automatically included or excluded from BPL status. The addition of the "no caste/tribe" answer option was not for these historically oppressed groups, which the survey sought to capture data on for the determination of BPL status. Relatedly, the government had been enumerating SCs and STs in all censuses in independent India. In fact, the caste question in Census 2011 (similar to previous censuses) was "is this person SC/ST?" As such, SC and ST communities have a long history of answering this question and providing specific caste names for decennial censuses, as well as other government surveys. The instruction manual for SEC survey enumerators includes detailed instructions on how to enumerate SCs and STs. After more than a page of directions, the manual states, "[I]f the respondent says that she/he does not belong to SC or ST then record '3' under this question against the name of that person."[27] The instructions clearly prioritize capturing SC and ST data accurately and then categorizing everyone else as "other," and are consistent with the enumerator manual for Census 2011. Second, the demands for including a "no caste/tribe" answer option did not emerge from groups seeking to have nationwide data on caste. Those organizing for a caste census overwhelmingly belonged to, or advocated on behalf of, caste-oppressed groups, particularly OBCs— who have been eligible for central government affirmative action benefits since the 1990s but have not been enumerated in a nationwide census. Members of caste-oppressed communities—including SCs, STs, and OBCs—more commonly understood why having detailed data on caste could help to make visible structural disadvantage and privilege and be used to improve and expand anti-caste and anti-poverty policies and programs.

The decision to include a "no caste/tribe" answer related to ongoing efforts to block a caste-wise enumeration in the census and beyond. A subset of public intellectuals in the Nehruvian liberal tradition believed that the government was violating long-standing ideals of democratic India by enumerating caste, as described in Chapter 3. Like the *meri jati Hindustani* campaign, this group saw the option of "no caste" as a progressive opt-out from the state's reinforcement of an identity through its enumeration in a nationwide survey. In addition, conservative Hindu nationalists also supported the inclusion of this answer option. For them, any question about caste divided Indians into categories that were irrelevant given their focus on strengthening Hindu political and cultural power. Both groups, with different politics, saw the inclusion of a "no caste" answer option as an opportunity to opt out from the government imposition of an irrelevant identity. This option appeased comparatively privileged groups who saw little benefit to having their caste identity enumerated and, thus, embedded an ideology of castelessness in SEC survey instrument. As such, the survey would document caste among historically oppressed groups while giving an opt-out that could be utilized by the comparatively privileged.

The SEC survey questionnaire also removed mechanisms in the pilot BPL survey that helped reduce the possibility that households within the same caste community might be classified differently (that is, due to minor variations in the spelling of a caste name or because some households might provide a general category [for example, a national-level administrative category] while others might respond with a local caste identity [for example, their *jati*]). In the pilot, the village survey included a list of all the castes in the community, while also identifying their broader administrative categorization (that is, ST, SC, OBC, or "general" category). Supervisors then post-coded the collected household-level caste-wise data to classify each caste by its associated administrative category in central and state government lists. The SEC survey did not involve the creation or use of caste lists, except for the use of state-level SC and ST lists; the enumerator did need to verify SC or ST identities by checking the caste or tribe name shared by the household against the state-specific list. For all castes categorized as "other" for the first caste question, the enumerator simply recorded whatever answer the respondent provided about caste—with no process of commensuration or standardization built into the survey. In fact, the enumerator manual welcomes a range of answers for those in the "other" category":

> It is difficult to define "Caste" or "Tribe" in the changing social situation. Persons may report the sub-caste, sub-tribe, clan name, sect, sub-sect, surname, title etc. as their caste/tribe's name. You should not attempt to classify or categorise. For the purposes of this question you have to write whatever name is reported by the respondent.[28]

This approach differed significantly from the pilot BPL which made every effort to ensure that the spelling and identification of each reported caste in household surveys matched with a caste identified in the village survey. The process of listing all castes as part of the village survey and then post-coding each caste in the household survey helped to standardize the caste-wise data at the village level—including systematizing spelling and minor variations of the same caste name. In contrast, the SEC survey answer options and instructions encouraged immense variations when documenting individuals belonging to the same caste, even within the same village and region. The decisions to include both the "no caste" and "other" answer options for the caste question (including not replacing the "other" category with the all-India administrative categories of OBC and "general"), and not create lists of local castes, was likely to compound the work involved in cleaning, standardizing, and tabulating the collected caste-wise data. While the creation of the caste lists would have increased the preparatory work related to the survey, local caste lists could have aided in processes of commensuration to enumerate a complex social hierarchy.[29] The impact of these decisions meant that the "no caste" option, when utilized, would most likely be an opt-out used by caste elites (since SC/ST status contributed toward BPL identification), while the post-enumeration categorization of castes that fell under all-India administrative categories of OBC and "general" would likely be very messy and cumbersome. The revisions made to the caste-related questions when finalizing the survey instrument supported an obfuscation of caste privilege and reinforced castelessness in the SEC survey.

Minimal Attention to Caste-wise Enumeration of Religious Minorities

The process of enumerating the caste of minority religious communities was complex—especially among religions whose Dalit members are not recognized as SC by the state. To do so effectively for the SEC survey would have required careful preparation, planning, field-testing, and revisions to the survey instrument, and similar attention on how to educate enumerators and supervisors through in-person trainings and written instructions.

Caste hierarchy exists across religions in South Asia—shaping patterns of marriage formation, employment, residence, dignity and respect, and access to places of worship. The relationship between caste and religion varies across regions, as well as historical patterns of migration, conversion, and colonialism. Despite the fact that neither Islam nor Christianity has a scriptural or theological basis for caste, research finds that both Dalit Muslims and Dalit Christians experience untouchability. Clean and unclean castes exist among Muslims, such that *ashraf* Muslims consider Dalit and "backward caste" Muslims as unclean and impure, and in some instances, Dalit Muslims have built their own mosques to avoid discrimination.[30] Yet Dalits categorized as SC by the government when Hindu—such as Dhobis and Lal Begis—are not categorized as SC when Muslim, despite similar experiences of untouchability and being treated as inferior. An estimated 80 percent of Muslims in India, or 10 percent of the country's population, are caste-oppressed.[31] Dalit Christians also experience untouchability, indignities, and discrimination and often worship in separate churches. They are severely underrepresented in religious orders and leadership positions, as well as in Catholic educational institutions.[32] An estimated 70 percent of the Christian population of India consists of Dalits.[33] Other divisions internal to religious communities such as sects or denominations coexist alongside caste in South Asia. For example, Shia and Sunni are two sects in Islam while Roman Catholic, Anglican, Baptist, and Methodist are denominations within Christianity. The SEC survey planners needed to understand this context and use it to shape the survey instrument and training materials for enumerators and supervisors to facilitate the successful enumeration of caste among religious minorities.

Reviewing the survey instrument, training materials, and enumerator and supervisor instructions, I found little evidence of attention to how caste and religion intersect, particularly for religious minorities. In addition, the SEC survey was likely to generate a wider set of responses to the question on religion as it was open-ended. The directive in the instruction manual for enumerators states that the enumerator should record whatever they are told for the question on religion:

> You are bound to record faithfully whatever religion is returned by the respondent for herself/himself and for other members in the household. You might come across situations wherein a person would like to mention the name of her/his sect, denomination, belief as the name of her/his religion. In such cases also, please do not attempt to classify the religion on your own but record faithfully whatever name of the religion that the person states.[34]

Having an open-ended question on religion in which enumerators recorded whatever answer the respondent provided was likely to create much greater variation in responses compared to previous surveys with categorical answer options, such as the pilot BPL or Census 2011.[35] While this approach allowed for a wider breadth of religions to enumerated it also meant that data would be more difficult to commensurate, and it would be challenging to anticipate how best to educate enumerators on the relationship between religion and caste prior to conducting interviews.

The manual and in-person training specified two main points about the intersection of religion and caste. First, the written and oral instructions conveyed that data on caste should be collected from every individual regardless of their answer to the question on religion. Second, the training materials for enumerators clearly stated how religion related to caste and tribe for SCs and STs. For the enumeration of SCs, the instructions and in-person trainings detailed the relationship between caste and Hinduism as well as other religions originating in South Asia. The instructions for enumerators were vague in describing the broader relationship between religion and caste, and particularly for religious minorities:

> In some religions, there may not be identifiable groups called "Castes/Tribes" but at the same time, there may exist identifiable groups known by different names that are equivalent to "Castes/Tribes." If the respondent says that she/he or any member of her/his household belongs to particular group and wishes to return the name of that particular group as her/his Caste/Tribe, in such a case, record "3" against that person in Col.16.[36]

The enumerator instructions and training emphasized the need to record whatever the respondent shared with regard to caste. While there was consistency in this approach, it assumed that collecting caste-wise data from all groups was equally straightforward. The silence on how to approach the enumeration of caste for Muslims and Christians left a void in the training. There was little attention to the challenges that might arise while trying to collect caste-wise data from religions that have no theological basis for caste but practically employ a caste system in South Asia. There was no distinction made between sect and caste for Muslims or between denomination and caste for Christians. The lack of clarity on how to enumerate caste for religious minorities—combined with an open-ended question on religion—greatly decreased the likelihood that the collected caste-wise data could speak to how caste operates as a system of domination among religious minorities. This negligence also aligns with a broader ideology of Hindutva which otherizes religious minorities—ignoring how caste-based privilege and discrimination operates within these communities.

The merging of the two projects meant that the SEC survey received an infusion of additional resources, despite the inattention to how caste and religion intersect for religious minorities, and the lack of widespread field-testing of the inserted caste-related questions. In previous BPL surveys and in the 2010 pilot survey, the government had collected household data using paper surveys. With the infusion of additional resources for the caste-wise enumeration, new options became available and led to the large-scale entry of the private sector.

Electronic Data Entry: The Entrance of the Private Sector

For the SEC survey, the use of tablet PCs during the household interview created the need to develop an infrastructure at the local level to collect and upload electronic data from each block.[37] Charge centers at the *tehsil* (rural) and ward (urban) levels continued to be the local hub for data collection operations where during past censuses enumerators and supervisors met to submit, review, and check paper records. Now they required an IT infrastructure to facilitate real-time electronic data entry during household interviews; this infrastructure consisted of of two PCs, a printer, a data backup system, charging centers for the handheld PCs, a generator, related furniture, and internet connectivity to upload the collected data to a central server at the National Informatics Centre (NIC). Three individuals were supposed to staff each charge center and prepare the tablet PCs for data collection (this would include a multistep process of preloading household data from the NPR), provide technical support to data collection teams if difficulties arose with the tablet or software program, and assist in downloading the collected data from the tablets.[38] Initially, the MoRD planned that Bharat Electronics Limited (BEL), a government-owned company or public sector undertaking (PSU) under the Ministry of Defence, would staff each charge center and hire and train contract workers for data entry during the household interviews.[39] By July 2011, BEL and two other PSUs took responsibility for overseeing the contracting out of electronic data-entry operations to the private sector—as will be discussed later in this section.

An entire human infrastructure accompanied the creation of the technological infrastructure to support "real-time" electronic data entry, starting with the individuals involved in collecting data from households. The planners now reimagined who would be involved in the household interview. Government schoolteachers were not available to serve as enumerators due to restrictions in the Right to Education Act, so state governments needed to identify a new cadre of government workers to staff the ground operations and canvass the survey. This group was expected to vary across locations. When the secretary of the MoRD in Delhi distributed plans for the SEC survey operations to the secretary of rural development for each state in late June 2011, the enumerators were described as being "appointed from among the revenue/ development/ health functionaries. Therefore, patwaries, panchayat secretaries, ASHA workers, anganwadi workers, municipal workers and postal workers may have to be appointed for this purpose."[40] In urban Karnataka, *anganwadi* workers, who serve in a variety of capacities as community health workers and daycare providers, formed an available cohort of workers.[41] Yet, in contrast to schoolteachers who minimally completed secondary school, pre-university, or had an undergraduate degree, *anganwadi* workers only required an 8th grade education. At the same time, they were usually knowledgeable about the local community and had prior experience serving as an interface between government programs and the community. The number of available *anganwadi* workers was generally insufficient to fill the required number of enumerators, so other government workers, including recent retirees, were also recruited as needed. Unlike for the decennial census, enumerators were asked to complete data collection in more than one block and usually four blocks in total.[42] A supervisor was appointed to check the fieldwork of every six enumerators.

The planners believed that *anganwadi* workers and other current and retired government employees were unlikely to have the technological literacy to enter household information into the data entry software on a tablet while conducting the interviews. As such, they reimagined the household operations with a two-person data collection team, including a government employee who would serve as the enumerator and a data entry operator (DEO) who would input information into the tablet. The census bureau would design and implement the training program for enumerators and supervisors, while the training of DEOs was initially under the domain of BEL and later contracted out to the private sector along with rest of the electronic data entry operations.[43] Planners envisioned that government employees serving as enumerators would operate as lead interviewers, while the DEO would have a limited technical role that would involve "operat[ing] the HHD [hand held device] in the field while the enumerator will ask the questions."[44] The DEO was envisioned in a secondary role to the enumerator during interviews.

By July of 2011, the central government announced that it had tasked the Consortium of Public Sector Undertakings (CPSUs)—consisting of BEL (who designed and built the tablet PCs and data-entry software), Electronic Corporation of India (ECIL), and Indian Telephone Industries (ITI) Limited—with implementing the electronic data collection throughout India under the supervision of the MoRD, MoHUPA, and ORGI.[45] The CPSUs then oversaw a tendering process to hire private companies to manage data entry within each district throughout the country. Previously, the private sector had limited involvement in nationwide censuses—with their input restricted to providing technology and software for data entry at the state or regional levels.[46] Now the scope of their presence increased. Companies were asked to submit proposals to run data-entry operations in specific states, from the hiring and training of DEOs to the setting up of charge centers.[47]

The CPSUs oversaw the process of advertising the "request for quotation" (RFQ) or scope of work for electronic data collection, processing, receiving, and rating proposals, selecting the tender for each state, and creating and monitoring compliance of the tender contracts. As per the application instructions in the RFQ, each tender had about two weeks to submit their bid from the date the RFQ was issued. In the case of ITI Limited, the PSU issued the bid on July 18, 2011, and vendor submissions were due on August 8, 2011. As part of the process, a pre-bid meeting was held on July 29, 2011. Vendors submitted bids to provide services for an entire state or multiple states and data collection was expected to take place between September and December 2011, with a staggered start across states.[48] In a relatively small state like Himachal Pradesh, with a population of seven million in 2011, this required collecting data across 12 districts and setting up 75 charge centers at the local level. In a state like Tamil Nadu, with a population of nearly 69 million people in 2011, the tender would be responsible for data collection in 31 districts and setting up 200 charge centers. The approved personnel for electronic data collection was 3,837 people for Himachal Pradesh and 35,879 for Tamil Nadu. The RFQ specified that "data entry should commence with 100% deployment of operators from day one" across the state and that "the maximum time limit is 45 days (including Sundays and holidays) for all states."[49] Only four large states (that is, West Bengal, Bihar, Maharashtra, and Uttar Pradesh) were permitted to stagger household interviews in two stages and had 90 days to complete data collection. The scale and speed at which the vendor needed to understand the project, submit the proposal, and (if selected) create a data collection infrastructure was rapid. The CPSUs reviewed applications and assigned tenders to states or portions of states (based on district boundaries) and expected data collection to begin immediately thereafter. In Karnataka, which is the focus of the next chapter, the SEC survey operations involved setting up 175 charge centers in the state's 27 districts to enumerate over 60.7 million people across 126,612 blocks.[50] Data collection for the project was projected to start in Karnataka in November 2011 and be completed in forty-five days.[51]

The decision to shift the project to real-time electronic data entry also led to a restriction in the language of data entry. Previously, answers were recorded on questionnaires translated into state/regional languages, and the enumerator could record answers in the same state/regional language. While it was possible to have electronic data entry occur in a variety of languages with unique alphabets, creating a program that allowed for that was expensive. The SEC data entry program allowed for data entry in English and Hindi.[52] In practice this meant that the language of the interview—including the language that the respondent shared answers in—and of data entry would often differ. To address the diversity of languages that interviews would be conducted in, the planners ensured that a paper version of the questionnaire and the training manual for enumerators and their supervisors would be available in the official state language as well as English or Hindi. In Karnataka, enumerators were expected to use a Kannada (or English) paper questionnaire to ask questions and then the DEO would record responses in English. It required that data collectors would be fluent in both languages, and throughout much of south India individuals who had completed 12th standard could read and write in English.

The envisioned timeline seemed ambitious and unrealistic given the extensive effort involved in setting up the data-entry infrastructure. The proposed schedule expected household interviews to begin between July and

December 2011, which meant that the preparatory work had to be finished already. Since not enough tablets were available for simultaneous data collection across the country, the plan included staggering data collection across states, such that tablets would be used to complete data collection in a particular state during a six–eight-week period and then be sent to another state. With a maximum timeline of forty-five days for data collection in each state (and ninety days in the four largest states), SEC survey administrators expected data collection to finish by February or March 2012, at the latest.[53] Yet the timeline did not appear realistic given that the SEC survey required hiring companies to set up the local electronic data-entry infrastructure; staffing the neighborhood-level charge centers; hiring and training master trainers, enumerators, and their supervisors; tablets arriving in a particular location on the scheduled date and then data-entry staff uploading block-level NPR data; and each data collection team administering the survey in up to four blocks. In contrast, enumerators had a month to complete all household interviews in a single block for Census 2011.

In a letter dated December 2, 2011, when data collection should have started in every state and been completed in several states, MoRD director N. K. Sahu writes about the disconnect between the imagined data collection and the emerging processes on the ground. After a meeting held between the MoRD and the CPSUs on November 19, 2011, Sahu documents the "delays in appointment of vendors by the CPSUs and the low wages being paid to DEOs" and states that if vendors failed to supply DEOs then state governments can either "take over the activities of such vendors and shall be entitled to reimbursement at the permissible rates" or "where no vendor has been appointed or where no DEOs have been appointed, the states may be permitted to take over the entire exercise."[54] Five months earlier, the chief economic officer of the MoRD, P. K. Padhy, had sent a letter announcing the start of the SEC survey in the state of Tripura on June 29, 2011 and had stated that the SEC survey would be completed by December 31, 2011; it now appeared that in many parts of the country, data collection might not even start by the original completion date for the project.[55]

The Envisioned Interview

The SEC survey administrators saw the work of enumerators as central to the completion of the household interview. The recruitment and training materials portray enumerators as essential to the project's success:

> You are privileged to have been chosen as an enumerator. As an enumerator, you are performing a duty which is of great national importance. While it is a matter of pride for you, it is at the same time, a great responsibility. You have to fill all tasks assigned to you with sincerity and devotion.[56]

Enumerators were expected to play the lead role in the household interview and were responsible for several other tasks, including completing the acknowledgment slip, updating the household listing and block map, and overseeing the work of the DEOs. The designers of the census, the regional managers of the census, and users of the census data envisioned the 600,000 enumerators as the backbone of census operations. As such, the training of enumerators was both broad in its scope and focused on numerous details pertaining to the process of interviewing households and the related documentation of interviews. In contrast, the DEO was envisioned to play a backend role as an electronic "stenographer"—entering what they heard into the tablet. Their short technical training focused on the use of the tablet PC and software program for data entry.

The planners of the SEC survey expected enumerators to begin the interview with a salutation and then explain the purpose of the interview:

> Good morning. I am <enumerator name>. Accompanying me, is <data entry operator name>. We are here to solicit your participation in helping us collect data for the Socio-Economic Caste Census. The information you give us will be compiled to help the government deliver benefit schemes more effectively and also formulate policies/schemes for Housing and Urban Poverty Alleviation. This Census is being conducted to collect socio-economic information related to the household. The information we collect will be displayed at various places in the Municipal corporation/council office and can be verified at the ward offices subsequently. We will also be issuing you an acknowledgement slip after collecting your information. Your honest cooperation in this process is solicited for this purpose.[57]

Following the brief introduction, the enumerator was supposed to show their identity card and collect information from an adult member of the household. The in-person training and training manual for enumerators explicitly advised them to not "assume any information" but "ask the respondent and note down only the information that she/he provides"; neither "prompt the respondent" nor ask "leading or suggestive questions"; "always be gender sensitive and ensure that no bias creeps in while canvassing the schedule"; follow the exact order of the questionnaire; if the respondent does not know the answer to a question, they "may be encouraged to consult with other knowledgeable members of the household."[58]

After the introductory remarks, the SEC survey planners envisioned that the enumerator would ask the respondent about the name of the head of household. If the respondent's answer matched the name recorded in the household listing created for Census 2011, then the enumerator would tell the DEO the unique number associated with the household in the list. The DEO was expected to enter the number into the data-entry program to seek a match in the preloaded NPR data. If the DEO was unable to find an electronic match for a particular household (because the household was not enumerated during the NPR, a new household now lived in the home, or a household lived in a place where previously there was no household), they created a new entry for the household and the enumerator also updated the household listing. If the DEO found a record for the household, then the first part of the interview would involve verifying the household's NPR data.

The enumerator then asked questions about each item on the survey—reading from a large 11-by-14 paper questionnaire included in the enumerator's toolkit. In urban areas, the enumerator first asked 16 questions about each member of the household: name, relationship to head of household, sex, year of birth, marital status, names of mother/father, occupation, highest educational level, main source of income/earnings, frequency of wages, disability, chronic illness, religion, and caste (2 questions). After posing these "individual level" questions, the enumerator asked 15 questions on "household particulars" that focused on the dwelling (that is, material of walls and roof, ownership status, and number of rooms), amenities (that is, drinking water, source of light, latrine, waste water outlet, separate kitchen), and household assets (that is, phone, computer, refrigerator, vehicle, AC, and washing machine). The rural questionnaire was similar but had fewer individual-level questions and additional questions about the household and land (see Appendix A).[59] The DEO was expected to record the household responses. The training manual for enumerators states, "[Y]ou must ask the questions, and ensure that the data entry operator enters the response into the hand-held device correctly."[60] Enumerators were expected to oversee the interview, ask questions, confirm answers, and clarify discrepancies.

After the enumerator collected individual- and household-level data, they needed to complete a few additional tasks. First, the enumerator was supposed to secure the respondent's permission to publicly post the collected data (except for the religion and caste data). Then, the enumerator and the DEO recorded whether they thought the collected information was accurate. The enumerator was also expected to complete an acknowledgment slip as proof that the interview had taken place. The enumerator and the DEO signed one portion of the slip and left it with the household, while the respondent signed the other portion of the slip which the enumerator returned to their supervisor.[61] Enumerators then put a sticker on the front door of each enumerated household and marked the front door with chalk to indicate that the house had been interviewed. After interviewing all the households in a block, the enumerator had to create an abstract summarizing key details about the block.

The collected BPL survey data, except for the caste and religion data, then underwent a period of public verification prior to finalizing the data. A summary of household-level data would be printed and posted publicly in a local government office, during which time anyone could review the entered data and register a complaint regarding errors or omissions with a local official. In addition, the government posted the data on a SEC survey website to allow anyone with internet access to review and file a complaint if they found errors in the data. This built-in system of verification ensured that mistakes made during data collection and entry could be corrected before the government used the data to determine eligibility for BPL status. However, since "information on the person's/household's religion and caste/tribe name [was not] published," the public had no opportunity to make corrections to the religion and caste/tribe data according to the planning documents circulated prior to the start of ground operations for the survey.[62]

Gaps in the Imagined Interview

The household interview appeared straightforward based on the planning documents and training materials.[63] Yet several noticeable gaps remained in the plans that would affect the collection of caste-wise data.

First, the ORGI did not have a recent template of how to train enumerators in the collection of caste-wise data, since it had neither collected caste-wise data during Census 2011 nor in previous censuses in independent India. The training materials used for the caste-related questions in the SEC survey were very similar to those used for the decennial census and insufficient. The instructions centered on the enumeration of SCs and STs, for which the decennial census had developed detailed instructions since Census 1951. No explanation was provided for how caste and religion might intersect for India's religious minorities—such as for Dalits, "backward" communities, or caste elites in the Muslim and Christian communities. The short explanation of the "no caste" answer option in the training materials remained silent on the motivations for including this answer option. However, colonial censuses had extensive experience collecting caste-wise data as discussed in Chapter 2. Across successive censuses in pre-independent India, regional administrators developed caste lists and specific instructions for enumerating caste. These historical documents could have provided a roadmap of how to enumerate caste in the SEC survey and the types of instructions necessary for successful data collection. The published caste tables from the 1921 and 1931 censuses also reveal the complex relationship between caste and religion, including attention to the presence of "untouchable" castes among Muslims and Christians, and could have afforded useful background information to enumerators and their supervisors.

Second, the "added-on" caste-related questions did not undergo extensive field-testing prior to the start of the survey. In addition, the new interview design with tablets prevented households from being able to view the recorded answers during the interview—unlike with past censuses. With paper records, households could see answers as interviewers documented them during the interview. They could point out mistakes during the interview itself. The public verification process helped address this issue for the BPL data, as it created an opportunity for households to request revisions to the draft data. However, the collected religion and caste-wise data were excluded in the public verification stage, so no built-in mechanism existed to correct these data.

The survey planners also imagined that households would be aware of the SEC survey at the time of the household interview. After an enumerator explained why the two-member data collection team was at the house, households would agree to be interviewed without hesitation. Yet the centralized planners delegated the burden of publicity to local government officials. Advertising the SEC survey became the duty of district-level officers, whose list of responsibilities included: "[G]ive proper publicity to SECC-2011 programs so as to get proper responses from the public."[64] Neither the state-level SEC survey principal officer to which each district officer reported nor the charge-level officers who reported to district officers had any publicity-related responsibilities in their list of roles and functions—as defined prior to the start of the SEC survey.[65] To expect each district officer to take on this role seemed unrealistic and at best would create a situation where publicity was uneven across districts and lacking in many places. In contrast, for the decennial census, the central government helped to roll out a massive campaign that played a key role in generating publicity for the project and in ensuring that households knew about the census prior to being interviewed. Now, the local officers who would be busy staffing ground operations, setting up charge centers, and overseeing ground operations also needed to develop a publicity campaign. Particularly in urban areas where previous rounds of the BPL survey had not occurred, it seemed unlikely that households would be aware of the SEC survey at the start of the project.

Finally, the training for the new DEO role was minimal. Neither the expertise accumulated by the ORGI over decades of conducting censuses and training enumerators, nor the knowledge that the MoRD had acquired during the pilot BPL shaped the training for DEOs. Private companies were responsible for hiring and training DEOs. Most DEOs received a short technical training on the tablet and had no knowledge of the purpose of the project nor specific question-and-answer options. Even if their role was envisioned as an electronic stenographer, the DEOs should have had knowledge about the survey content.

Looking Back and Ahead

The revisions to the BPL survey, which were piloted in September 2010, sought to improve the major shortcomings of past rounds of the survey as identified by the Saxena Committee (that is, the expert group tasked to improve the BPL methodology):

> In summary exclusion errors from BPL lists frequently reflect the powerlessness of the most vulnerable and are a direct function of their weak political bargaining power as a collective entity in a democracy. Our inability to include them in State programs in the last sixty years is a severe indictment of public policy and its implementation. To address this, the Expert Group, therefore felt it was imperative that the foremost task remained that of ensuring their automatic inclusion as clearly defined social and economic categories.[66]

One of the recommended changes included collecting caste-wise data at the household level and post-coding this data with associated central and state government administrative categories based on the listing of castes created by the field supervisor in the village survey. Yet, once the political leadership chose to combine the BPL survey and "caste census," the caste-related questions from the pilot BPL were replaced with a set of question and answer options that had not undergone extensive field-testing and more closely resembled recent decennial censuses. Processes of bureaucratic deflection at this stage involved backtracking on revisions to the piloted BPL survey and modifying the caste-related questions to include answer options that could help to obscure caste-based power; the development of an unrealistic timeline for completing data entry; not spending resources on centralized advertising for the survey; and not creating state-specific caste lists prior to collecting data that could have aided in processes of commensuration.

The wording of the caste-related questions did not emerge from an anti-casteist ideology. The inclusion of "other" as an answer option instead of the administrative categories of "OBC" and "general," and the insertion of the "no caste" answer option, embedded castelessness in the survey. In addition, the centrally developed enumerator training failed to discuss the relationship between caste and religion—particularly for Muslims and Christians. Given that over 200 million Indians are Muslims, this was a significant omission, but one that was in line with an ideology of Hindutva. Once stripped of anti-casteist politics, the caste-wise enumeration took on ways of seeing and operating in line with existing power hierarchies. As Melissa Nobles argues, "the real power of census methods, however, has been to present the process of categorization as a technical procedure, and not a political decision."[67]

The decision to utilize real-time electronic data entry required scaling up the on-the-ground infrastructure to make this possible. The government and the media heralded the inclusion of tablets as a sign of progress and development and focused on this decision as evidence that the exercise was well designed. In reality, the use of tablets diverted attention away from two important areas that required a significant investment of time and resources. First, there was no effort to develop a centralized publicity campaign to inform the general public about the project. Households had limited knowledge about the project and its purpose—particularly in urban areas where BPL surveys had not been previously conducted. Second, there was no attempt to expand the state-level SC and ST lists to create a comprehensive list of castes in each region. This time-consuming but important step prior to interviewing households would have significantly streamlined the process of collecting and analyzing the caste-wise data.

With the start of the revised BPL survey imminent, one expert involved in the centralized effort to design and implement the survey explained:

> What actually happens in the field remains a large mystery. We have all these procedures on paper, but it's completely different to implement them in practice. For example, will people understand the questions that are being asked? And,

> when they are given a receipt at the end of the interview, will they understand and believe that what they have said has been captured? Unlike paper schedules, they will not be able to see the recorded answers.[68]

This expert expressed concerns that respondents might not understand the purpose of the data collection or the questions asked, or be able to check whether the recorded data captured their answers, since they would not be able to see what the DEO entered into the tablet. At the same time, the main purpose of the exercise remained to identify poor households. This was obvious from the multiple checks to ensure the questions and answer options for the BPL portion of the survey were revised based on the Saxena Committee recommendations and extensively field-tested, and that the collected data were publicly verified, unlike the caste-wise portion of the survey. The next chapter focuses on ground operations of the survey based on fieldwork in the south Indian state of Karnataka to explore the process of producing caste-wise data during the household interviews.

Notes

1. M. Vijayanunni, "BPL Surveys Are Mere BPL Surveys, Not Caste Census: Dr M Vijayanunni," *Economic Times*, July 31, 2011, https://economictimes.indiatimes.com/opinion/interviews/bpl-surveys-are-mere-bpl-surveys-not-caste-census-dr-m-vijayanunni/articleshow/9423430.cms, accessed April 1, 2024.
2. Government of India (N. C. Saxena, Chair), *Report of the Expert Group to Advise the Ministry of Rural Development on the Methodology for Conducting the Below Poverty Line (BPL) Census for the 11th Five-Year Plan* (Krishi Bhavan, New Delhi: Government of India, Ministry of Rural Development, August 2009), https://rural.nic.in/sites/default/files/ReportofExpertGroupChaired-Dr.N.C.Saxena.pdf, accessed April 1, 2024.
3. Bureaucratic deflection refers to the process by which the administrative bureaucracy silently maneuvers, avoids, and resists the implementation of policies and programs that challenge Brahmanical patriarchy and Hindutva ideologies, as defined in Chapter 1 and elaborated in Chapters 2 and 3.
4. Castelessness has strengthened as an ideology since the 1930s as caste-elite Congress Party leaders muted their caste identities to become credible *all-India* representatives, in response to oppressed-castes' assertions for equality, dignity, and the dismantling of caste hierarchy. S. Deshpande, "Caste and Castelessness."
5. Government of India (N. C. Saxena, Chair), *Report of the Expert Group to Advise the Ministry of Rural Development on the Methodology for Conducting the Below Poverty Line (BPL) Census for the 11th Five-Year Plan.*

6. Ibid., 23.
7. The eight criteria for automatic inclusion of households were (*a*) "primitive tribal groups"; (*b*) the most discriminated against SC groups called Mahadalits; (*c*) single women headed households; (*d*) households with a disabled person as bread-earner; (*e*) households headed by a minor; (*f*) destitute households which are dependent on "alms" for survival; (*g*) homeless households; (*h*) if any member of the household is a bonded laborer. Ibid., 26.
8. Ibid., 21.
9. The sample for the pilot BLP was drawn from the 66th round of the National Sample Survey. Press Information Bureau, Government of India (Ministry of Rural Development), "Below Poverty Line (BPL) Census to Be Conducted along with Caste Census during June–December 2011."
10. For each group, the supervisor manual instructed them to "write the religion code the caste practices" in the first column of the table, and then record "the name of the caste or tribe" in the second column. The village survey conceptualized religion with six specified options (that is, Hinduism, Islam, Christianity, Sikhism, Jainism, Buddhism) along with the seventh option of "other." Ministry of Rural Development, "Socio-Economic Survey 2010: Supervisor Manual" (Government of India, New Delhi, 2010), 59.
11. The category of "Other Backward Caste (OBC)" is listed in the English-language version of the village schedule for the pilot BPL; however, the standard government administrative category is "Other Backward Classes." Ministry of Rural Development, "Socio-Economic Survey 2010: Village Schedule" (Ministry of Rural Development, New Delhi, 2010), 4.
12. The sample village survey listed 1 = Maha Dalit (poorest Dalit communities), 2 = Most Backward Caste, 3 = Backward Caste; 4 = Primitive Tribal Groups; 5 = De-notified Tribes; 97 = Other (specify).
13. The Saxena Committee recognized that caste privilege structured socioeconomic privilege. The revised BPL survey tried to address the problem of past surveys that had mistakenly identified comparatively privileged households as BPL—largely due to social networks and connections that assigned them to BPL status. The Saxena Committee recommended that "some of the most transparently powerful categories of rural households—economically, socially and politically—be automatically excluded, so they would not be surveyed, and thus will not be able to edge their way into the BPL lists." Government of India (N. C. Saxena, Chair), *Report of the Expert Group to Advise the Ministry of Rural Development on the Methodology for Conducting the Below Poverty Line (BPL) Census for the 11th Five-Year Plan*, 21.
14. The BPL administrators designed and administered the village and household surveys such that the answers to caste and religion questions in the household surveys were then associated with an entry in the village survey.

15. Ministry of Rural Development, "Socio-Economic Survey 2010: Supervisor Manual," 19.
16. Ibid.
17. Ibid., 19–20.
18. Ibid.
19. Ibid.
20. Ibid.
21. Ministry of Rural Development, Government of India, "Guidelines to States for Conducting Pilot Socio-Economic Survey 2010," July 2010, https://rural.gov.in/sites/default/files/Guidelines_4_States.pdf, accessed April 1, 2024.
22. The first three answer options were identical to those in Census 2011.
23. Similar to the decennial census, state-specific caste/tribe categories were listed as answer options in the second caste question for those who identified as SC or ST.
24. Household-level data played a central role in both automatically including and automatically excluding households. The five criteria to automatically include households for BPL status shifted somewhat from the Saxena Committee's recommendations but remained consistent before and after the pilot BPL and SEC survey: (*a*) homelessness, (*b*) if the primary source of income is from begging, charity, or alms, (*c*) if the household includes manual scavengers, (*d*) if the household includes members of primitive tribal groups, and (*e*) if the household includes legally released bonded laborers. Similarly, any household satisfying one (or more) of the following 14 criteria from the household section of the questionnaire would be automatically excluded from BPL status: (*a*) motorized two/three/four-wheeler/fishing boat, (*b*) mechanized three/four-wheeler agricultural equipment, (*c*) Kisan Credit Card with credit limit of INR 50,000 and above, (*d*) any member as a government employee, (*e*) non-agricultural enterprises registered with the government (*f*) any member of the family earning more than INR 10,000 per month, (*g*) pay income tax, (*h*) pay professional tax, (*i*) three or more rooms with all rooms having *pucca* walls and roof, (*j*) own a refrigerator, (*k*) own landline phone (*l*) own 2.5 acres or more of irrigated land with at least one irrigation equipment, (*m*) 5 acres or more of irrigated land for two or more crop seasons, (*n*) own at least 7.5 acres of land or more with at least one irrigation equipment. Ministry of Rural Development, "Socio Economic & Caste Census 2011 in Rural India" (Government of India, July 25, 2011), 25–26, https://rural.gov.in/sites/default/files/Caste%20Census%20book%2017-11-2011_0.pdf, accessed April 1, 2024.
25. Ministry of Housing and Urban Poverty Alleviation, "Instruction Manual for Enumerators, Socio-Economic and Caste Census 2011-Urban" (Government of India, 2011), 47, https://mohua.gov.in/upload/uploadfiles/files/31375(1).pdf, accessed April 1, 2024.

26. Ibid., 48.
27. Ibid., 48.
28. Ibid., 50.
29. The caste lists could have aided enumerators during household interviews (similar to state-level SC/ST lists) and/or allowed for post-coding of caste data by local supervisors (as occurred in the pilot BLP).
30. Imtiaz Ahmad, *Caste and Social Stratification among the Muslims* (Delhi: Manohar Book Service, 1973); Imtiaz Ahmad and Shashi Bhushan Upadhhyay, eds., *Dalit Assertion in Society, Literature and History* (Delhi: Publisher Orient Blackswan, 2010); Irfan Ahmad, "A Different Jihad: Dalit Muslims' Challenge to Ashraf Hegemony," *Economic and Political Weekly* 38, no. 46 (2003): 4886–91; Ghaus Ansari, *Muslim Caste in Uttar Pradesh: A Study in Culture Contact* (Lucknow: Ethnographic and Folk Culture Society, 1960); Prashant K. Trivedi, Srinivas Goli, Fahimuddin, and Surinder Kumar, "Does Untouchablity Exist among Muslims: Evidence from Uttar Pradesh," *Economic and Political Weekly* 51, no. 15 (April 9, 2016): 32–36.
31. Rajesh Kumar Nayak, "Quest for Equality and Identity: The 'Pasmanda' Muslim Discourse in Post-1947 Bihar," *Proceedings of the Indian History Congress* 74 (2013): 961–71.
32. Prakash Louis, "Dalit Christians: Betrayed by State and Church," *Economic and Political Weekly* 42, no. 16 (2007): 1410–14.
33. Ibid. Christians make up approximately 2 percent of the Indian population.
34. Ministry of Housing and Urban Poverty Alleviation, "Instruction Manual for Enumerators, Socio-Economic and Caste Census 2011–Urban," 45.
35. The census did allow for a broader range of responses to the religion question, despite the categorical answer options, as the final option was "other" and if selected the enumerator could enter the name for another religion.
36. Ministry of Housing and Urban Poverty Alleviation, "Instruction Manual for Enumerators, Socio-Economic and Caste Census 2011–Urban," 48.
37. Consortium of Public Sector Undertakings (CPSUs) Tender Processing PSU ITI Ltd., "Socio-Economic Caste Census-2011, Request for Quotation: Service Contract for Data Collection on Door to Door Basis for 120 Crore Population," July 18, 2011; Sanjay Kumar Rakesh, Ministry of Rural Development, "Infrastructure Requirements for Charge Centres Data Center (Tehsil Level) for SECC 2011," Do. No. 14016/6/2011AI (RD), July 11, 2011, https://rural.gov.in/sites/default/files/Letter%20dated%2011.7.2011%20Regarding%20Infrastructure%20requirements%20for%20Charge%20Centres_0.pdf, accessed April 1, 2024.
38. CPSUs Tender Processing PSU ITI Ltd., "Socio-Economic Caste Census-2011, Request for Quotation."
39. B. K. Sinha, Ministry of Rural Development, "Socio Economic Caste Census Guidelines (with Attachments)," No.Q-14015/04/2011 AI (RD), June 24,

2011, https://rural.gov.in/sites/default/files/Guidlines_27062011_0.pdf, accessed April 1, 2024.

40. B. K. Sinha, Ministry of Rural Development, "Socio Economic Caste Census Guidelines (with Attachments)."
41. *Anganwadi* workers serve children and families throughout India to combat hunger and malnutrition, provide access to immunizations, offer contraceptive counseling and access, arrange for health checkups and referrals, and run daycare services. Many of their services are offered in conjunction with India's public health care system. Depending on their appointment (that is, as a "helper" or a "worker"), *anganwadi* workers must have completed the 8th standard or the 10th standard.
42. Ibid.
43. Ibid.
44. Ibid., 3.
45. CPSUs Tender Processing PSU ITI Ltd., "Socio-Economic Caste Census-2011, Request for Quotation."
46. When surveys involved the use of paper records, electronic data entry usually took place at the state level. In the case of Census 2011, census officials had a robust computer lab in each state-level census directorate, where they received and accounted for all records, processed them through a software program that recognized handwriting and converted the written data into electronic form, and then manually checked cases flagged by the software program where "reading" the handwriting and electronic conversion were difficult. In Census 2011, a private company produced the software program that scanned and electronically "read" the paper surveys and converted the written text into electronic text. This important step in commensurating data was decentralized to the state level and aligned with the concentration of linguistic and cultural knowledge. Interview with government employee, July 22, 2011.
47. CPSUs Tender Processing PSU ITI Ltd., "Socio-Economic Caste Census-2011, Request for Quotation."
48. Ibid. Vendors submitted one proposal that included bids for either one or multiple states (that is, up to ten states) and ranked their preference of states.
49. Ibid., 39.
50. Each enumeration block contained 125 to 150 households. Ibid., 20, 33, 38.
51. P. K. Banerjee, ORGI, Ministry Home Affairs, "State-Wise Number of EBs/ Sub-Blocs for Use in the SECC," No. 32/4/20-CD (CEN), November 8, 2011.
52. B. K. Sinha, Ministry of Rural Development, "Socio Economic Caste Census Guidelines (with Attachments)."
53. CPSUs Tender Processing PSU ITI Ltd., "Socio-Economic Caste Census-2011, Request for Quotation," 38.

54. N. K. Shau, Ministry of Rural Development, "Letter to Primary Secretary of All State Governments and Nodal Officers, RGI, HUPA, and CPSUs," Doc. No. A-16015/07/2011-AI (RD), December 2, 2011.
55. P. K. Padhy, Ministry of Rural Development, "Instructions for Opening a Dedicated Account for the Socio-Economic Survey-2011," Doc. No. A 16025/8/2006-AI (RD), July 1, 2011, https://rural.gov.in/sites/default/files/Guidelines%20Norms%20for%20Utilization%20of%20Funds%20for%20conducting%20SECC%202011_1.pdf, accessed April 1, 2024. The secretary of the MoRD communicated a similar timeline in a letter sent to chief secretaries of every state in late May 2011. B. K. Sinha, Ministry of Rural Development, "Socio Economic Census to Identify Below Poverty Line (BPL) Families," No.Q-14016/6/2011 AI-(RD), May 30, 2011.
56. Ministry of Housing and Urban Poverty Alleviation, Government of India, "Instruction Manual for Enumerators, Socio-Economic and Caste Census 2011-Urban," 1.
57. Ibid., 3.
58. Ibid.
59. The SEC survey consisted of 13 individual-level questions for rural households and 23 questions at the household level. The second section of the "household particulars" included three yes/no questions (Q 5–7) about whether any member of the household was (5) from a primitive tribal group? (6) a legally released bonded laborer? (7) a manual scavenger?
60. Ministry of Housing and Urban Poverty Alleviation, Government of India, "Instruction Manual for Enumerators, Socio-Economic and Caste Census 2011-Urban," 15.
61. Enumerators completed a receipt slip for each household. The enumerator filled in basic information including the name and gender of each member of the household, the total number of individuals in the household, and identifying information about the enumerator and the DEO.
62. B. K. Sinha, Ministry of Rural Development, "Socio Economic Caste Census Guidelines (with Attachments)."
63. The ORGI's assigned role as technical experts for the SEC survey resulted in the training materials for the decennial census becoming the foundation for the enumerator training materials in the SEC survey.
64. B. K. Sinha, Ministry of Rural Development, "Socio Economic Caste Census Guidelines (with Attachments)," 17.
65. Ibid., 16, 18.
66. Government of India (N. C. Saxena, Chair), *Report of the Expert Group to Advise the Ministry of Rural Development on the Methodology for Conducting the Below Poverty Line (BPL) Census for the 11th Five-Year Plan*, 23
67. Nobles, *Shades of Citizenship*, 170.
68. Field notes, July 28, 2011.

5
The Household Interview

We are standing in front of a three-floor apartment building in central Bengaluru. "Thirty-seven," says Vijaya as she locates the building on the block map and shares the number from the map associated with the building. She folds the map and places it under the household listing booklet on her clipboard. Vijaya has been trained as an enumerator for the SEC survey and she and her husband, Mohan, who is the DEO, are getting ready to interview their first household of the afternoon. She opens the household listing booklet and flips through it until she reaches the three entries for building number 37. She locates the number associated with the ground floor unit and tells her husband, "148." Mohan types the number into his handheld tablet. The tablet contains preloaded data from the NPR, a recently created government database for a system of national identification. A few seconds later the data for household 148 populates the screen. Mohan reads aloud the name of the head of household, "R.L. Suresh." Vijaya nods to indicate that "R.L. Suresh" matches the handwritten entry for the head of household in the household listing booklet, which census enumerators created during the first round of Census 2011 two years earlier. Mohan opens the front gate, and we walk into the small compound and up to the front door of the ground-floor unit. Mohan knocks on the front door with his right hand and holds the tablet in his left hand. A few seconds later, a woman in her mid- to late thirties opens the door and looks at the three of us. Mohan holds up the handheld device as he explains that we are here "for census work."[1] He then asks, "Is this the house of R.L. Suresh?" The woman nods in acknowledgment. She invites us into the drawing room and asks us to sit. As we settle into chairs, she closes the front door and stands near the entrance. Her teenage son stands in a corner of the room and watches. Mohan confirms the name, sex, and age of the four household members based on preloaded NPR data.[2]

After verifying basic demographic information about each member of the household (that is, the respondent's husband, the respondent, their son, and the respondent's mother-in-law), Mohan asks the respondent each subsequent question that appears on the screen. He first asks individual-level questions about the head of household and enters their data, before asking the same questions for subsequent members of the family. Mohan begins with the husband of the respondent and inquires about his level of education and occupation. The woman responds "degree" and "works in a bank." For the question on "highest education level completed," Mohan selects "7: graduate or higher" and then chooses "9: other work" for the question on "main source of income." He also enters data for the subsequent question on "frequency of wage payments" and selects "monthly." He makes this inference about the occurrence of wages based on the husband's employment in the formal sector. Mohan then asks the woman if there is anyone in the household who has a disability. The woman mentions someone has "sugar," and Mohan qualifies "by birth." The woman shakes her head, indicating no. He records "7: non-disabled" for the disability question and "6: no chronic illness" for the subsequent question on chronic illnesses. Later he records the same responses for the two health-related questions for the three remaining members of the household. By now Vijaya has filled out and signed the SEC survey acknowledgment slip, which lists the names of each member of the household. She only requires a signature from the respondent and will ask for it when Mohan has completed the interview.

A few minutes into the interview, the woman asks us if we would like something to drink. We decline; it's a temperate afternoon in February and this is our first interview of the afternoon. Vijaya and Mohan have three to four hours of work ahead of them. I have been accompanying them for a week and will continue to do so for another week, as they enumerate four blocks in the center of the city of Bengaluru in early 2012.[3] Mohan continues to ask questions about the head of household and enters responses into the tablet. He asks the respondent, "Caste?" At the same time, he enters "Hindu" for the religion of head of household, without directly asking the woman about her husband's religion. The woman responds "Sri-xxxxxx." Mohan looks up briefly—signaling that the woman should repeat herself. The woman says what sounds like "Srivaishna," though she speaks quickly and quietly. Mohan continues to look at the woman. After another moment's pause, she explains, "Brahman." Mohan nods and looks back down at his computer screen. He selects "3-Other" for the categorical question on caste, and then types

"Brahman" for the subsequent open-ended question on caste. Later, when he enters the data for the other three members of the household, he enters "Hindu" for religion, selects "3-Other" for caste, and types "Brahman" without asking. Vijaya observes the back and forth between Mohan and the woman while organizing the documents on her clipboard. Once the interview is complete, Vijaya asks the respondent to sign one side of the perforated acknowledgment form, which Vijaya keeps; she gives the other half of the form, which she has already signed, to the respondent as proof of the interview.

Starting in December 2011, data collection teams like Vijaya and Mohan completed household interviews for the SEC survey throughout the south Indian state of Karnataka. The central planners of the survey imagined a two-person data collection team, consisting of an enumerator and a DEO, interviewing all households in four census blocks over the course of a month. Enumerators, like Vijaya, were usually government employees who received a substantial payment to conduct the interviews (especially in comparison to their monthly salaries). DEOs, like Mohan, were private citizens, hired by a firm that had secured the government contract to oversee data entry operations in Bangalore Urban district. Every afternoon, Vijaya left her job at the government daycare/nursery, and her husband closed his small print shop during the afternoon lull, to complete interviews. DEOs were sometimes, but not always, family, friends, or known acquaintances of the enumerator in Bengaluru.

This opening vignette highlights several differences between the planned interview and actual interviews for the SEC survey. In Chapter 4, I describe how government administrators in Delhi involved in the design of the SEC survey imagined the household interview. In this chapter, I draw upon observations of actual household interviews and ground-level survey operations in Bengaluru to trace the compounding effects of earlier decisions on the production of caste-wise data. I conducted fieldwork related to this chapter during five trips to Bengaluru ranging from three weeks to two months in length between 2011 and 2013. I describe the methodology in greater detail in Appendix A. Although Vijaya was hired as an enumerator, she played a minimal role in asking questions and collecting answers during interviews. The planners of the SEC survey envisioned that enumerators would use their expertise to conduct the interview, but Vijaya's role was instead limited to non-interview tasks. She identified each household using the block map and household listing and (if she found a match) conveyed a unique identifier to Mohan, so that he could retrieve and verify the household's pre-populated NPR data. Vijaya completed an acknowledgment slip for proof of interview and afterwards placed a sticker on the front door of the house to mark that the SEC survey interview had taken place. She also updated the household listing for each block (for example, if a new household resided in a home previously enumerated for the census or if she and Mohan located a new place of residence within the block). While these non-interview-related tasks were important book-keeping responsibilities, they did not take advantage of Vijaya's sociological sensibilities during the interview. She rarely utilized the information about the survey questions and interviewing skills that she learned during the mandatory two-day enumerator training developed by the ORGI, or from her profession as an *anganwadi* worker at a nursery. Instead, Mohan became the de facto enumerator; he asked questions, recorded the respondent's answers, and his understanding filled the space for interpretation and judgment during the interview. He made assumptions based on what he saw, and in certain cases he recorded answers to particular questions based on his observations without confirming details with the interviewee. While gender norms influenced Vijaya and Mohan's interview dynamics, as did their personal relationship as a married couple, I also observed a similar pattern across interview teams that entered data directly into the tablet. DEOs had much more power than imagined in every instance that I observed a two-person data collection team interviewing households and recording responses into the software program for real-time electronic data entry. This observation contrasted sharply with the expectations of project administrators in Delhi and folks overseeing ground operations in Bengaluru, who envisioned DEOs as electronic stenographers who would simply record what enumerators told them.

Mohan's understandings of religion, caste, gender, and family shaped the collection of data during the interview. He recorded that each member of the household was "Hindu" without explicitly asking about religion. The names of family members, religious icons, and items in and outside the home, as well as the reply to the question on caste, provided Mohan with sufficient details to feel confident when he recorded the religion of each family member without asking.[4] Mohan inquired about the caste of the head of household and unintentionally shaped the final answer to this question with his nonverbal cues. The respondent shifted from answering with a local caste name to a caste identity that has worldwide resonance (that is, Brahman). While Mohan played a role in reclassifying this interviewee's response, and in doing so further standardized the data, Mohan's reclassification of local castes as

"Brahman" was uneven. Earlier in the week, I observed him record a Brahman subcaste as "Iyer"—as shared by a respondent.[5] Mohan also attributed the caste of the male head of household to all other members of the family. Mohan's assumption that the head of household's identity applied to his wife, child, and mother reinforced Brahmanical and patriarchal norms and beliefs. In general, Mohan's role in classifying people by caste and religion led to an increased simplification of the data—like categorizing all family members as "Brahman" and attributing the caste and religion of the head of household to other family members.

This chapter examines how the process of collecting data from households during the SEC survey unfolded between December 2011 and early 2013. While the central government planned that the SEC survey would take six months to complete, the project ended up spanning several years. I first explore how decisions (and omissions) by centralized planners shaped processes of enumeration in Bengaluru, and then focus on how the enumeration and classification of caste-wise data reproduced inequalities. Based on observations of SEC ground operations, I show how ideologies of castelessness, Hindutva, and patriarchy heavily structured the production of caste and religion data.

The Infrastructure of Enumeration

An interdisciplinary body of scholarship by historians, anthropologists, and other social scientists explores the relationship between nationalism, state formation, and processes of enumeration.[6] Enumeration refers to the practices, social relations, and objects involved in the systematic counting of things. Processes of enumeration involve considerable work by local actors to build consent for participation in censuses and surveys. For in-person surveys, like the SEC survey, enumeration centers on the exchange between data collectors and respondents but is also structured by earlier decisions made by centralized project planners, as well as regional and local officials involved in overseeing the project's implementation and ground operations. Enumeration focuses on the social process of producing data during the household interview but is also heavily shaped by the survey instrument, the training of enumerators, and the tools and technologies of data collection that identify households and create legitimacy for the project (for example, the tablet PC, the electronic data entry program, the preloaded NPR data, and the household listing).[7]

The designers of the SEC survey envisioned that processes of enumeration would unfold in a particular way, as detailed in Chapter 4. Several decisions by centralized planners impacted processes of enumeration, from the decision to use tablets during the household interview to the follow-up plan to create a two-person data collection team. In addition, the content of the survey, such as the "other" and "no caste" answer options, and the structure of the data entry program shaped enumeration processes. The very different trainings developed for enumerators and DEOs impacted the interviewers' knowledge of the project. In addition, limited advertising prior to and during the project influenced the respondents' lack of prior awareness of the survey and eagerness to participate.

Since enumeration requires the input from the soon-to-be enumerated, respondents must be willing to share information about themselves and other members of their household; that is, they must consent to the process. Much of the work of engaging households to participate occurs by individuals either marginally connected to the state or outside of the state. In the case of the SEC survey, I find that local data collectors did much of the *consent-building* to ensure participation.[8] Several key decisions made by centralized administrators shaped consent-building and led to enumeration processes that differed from the original plans.

The Invisible Survey

In Bengaluru, like cities throughout Karnataka, negligible advertising by central, state, and local governments meant that households did not know that a statewide enumeration was underway at the time the SEC survey began in late 2011.

The Department of Rural Development and Panchayat Raj (RDPR) in Karnataka oversaw the implementation of the SEC survey. It continued the pattern of decentralization started by the MoRD offices in Delhi and gave money to each district within the state to publicize the survey.[9] When the enumeration began in parts of Bengaluru in late November and early December of 2011, there was virtually no government-sponsored publicity related to the project. In a city where local leaders plastered life-size advertisements in public locations to announce the birthday of a local MLA (member of legislative assembly) or a state holiday, there were no public announcements, billboards, newspaper ads, or public information campaigns throughout the city or in specific neighborhoods. Television ads were also non-existent until late in the data collection operations, when a sporadic TV ad ran on local stations in Bangalore Urban district.[10] At the start of data collection, the SEC survey was invisible to the public.

Media coverage was also limited in the lead-up to and during the start of data collection. Between September and November 2011, just one article mentioned the project in *Vijaya Karnataka*, the Kannada newspaper with the largest circulation.[11] Local editions of the three leading English-language newspapers in the state's major cities had sporadic coverage in the period leading up to the canvassing of the survey in Karnataka.[12] The relative silence around the project meant that households lacked independent knowledge about the questions and answers options, and did not understand the survey's purpose at the start of data collection.

During my fieldwork in the periphery of Bengaluru at the start of survey operations in December 2011, interview teams repeatedly conveyed that the households they visited were not aware of the SEC survey. During one of these conversations at a local charge center—or the neighborhood office set up to coordinate data collection operations—a small truck carrying a large speaker drove by an open window as it wove up and down the surrounding lanes to advertise a government health ID card. One enumerator pointed out, "This is what we need! If they would just make a similar announcement about the SEC Census then we wouldn't have to explain to each household."[13] As this conversation suggests, the lack of publicity surrounding the SEC survey left data collectors to be the primary, and in most cases the only, line of communication about the purpose of the survey. In contrast, the central government poured money into advertising the decennial census, with large billboards, radio, and TV ads, made additions to primary and secondary school curriculums, encouraged media coverage, and supported a host of other advertisement and educational efforts in advance of, and throughout, the decennial census operations. During the heart of the SEC survey operations in Bengaluru, government-organized publicity was negligible, and media coverage remained limited.[14] In addition, organizing by caste-based associations was largely absent, as their knowledge of the project before it rolled out was limited due to the minimal local publicity.[15] Households knew neither that the SEC survey was distinct from the recently completed Census 2011 nor that it was a revamped version of the rural BPL survey. The burden of consent-building was greater on interview teams for the SEC survey compared to Census 2011 due to the lack of state publicity and local media coverage.

The Mistaken Census

The lack of publicity surrounding the SEC survey, and its proximity to the well-advertised decennial census, led many urban households to believe the "census" referred to by data collectors was in fact the decennial census.

From the opening moments of the interview in the introductory vignette, Mohan used the presence of the tablet, preloaded NPR data, and a vague reference to "the census" to build trust and rapport with the household. Mohan's approach communicated prior knowledge of the household and created legitimacy by connecting the data collection team to the well-known and widely respected decennial census. Like Mohan, DEOs regularly displayed their tablets at the start of the interview to signal the "officialness" of their work and made references to "the census" to communicate the importance of the project and ensure smooth participation. The tablet helped build rapport, and DEOs used this technology to convey authority and fend off questions about the purpose of the project. As a result of cues from data collectors, and the lack of publicity for the survey, households in Bengaluru commonly assumed that the interview was related to the decennial census. This "mistake" was particularly understandable given the survey's proximity to the decennial census.

While the observed roles of enumerators and DEOs unfolded in unexpected ways in India's third-largest city, the patterns were neither random nor nonsensical. They emerged to fill the gap between information that the centralized administrators planned for and what was required to ensure participation within the context of local social norms.

The Elevated DEO and the Relegated Enumerator

The centralized planners and regional managers of the SEC survey envisioned enumerators as the backbone of ground operations. During an enumerator training in Bengaluru, a master trainer explained to a room full of future enumerators, "You are to ask the questions and then tell the DEO what to enter based on the respondents' replies."[16] However, as described in the opening vignette, enumerators often took a backseat role during household interviews.

In urban Karnataka, local governments recruited *anganwadi* workers, like Vijaya, as well as other government employees for enumeration work. The remuneration of SEC survey staff, training materials, and government notifications all depicted the enumerator in a crucial role and their work as being of national importance. One enumerator, who was a forest service employee, explained to me at a local data-processing center during his first day of interviewing households:

> I have worked for the government for more than twenty years. When I heard that the government needed enumerators for this census, I thought I should

> serve my government. My supervisor was also willing to let me leave work early for the duration of the census work. This work is important for the government and as a government worker I must help.[17]

The enumerator's role was understood to have a larger purpose, and they received a significant stipend for this work. At the time, an *anganwadi* worker in Bengaluru usually received a monthly salary of INR 6,000–8,000 (USD 90–130). In contrast, they received a total payment of INR 18,450 for enumerating four blocks.[18] At more than double and potentially triple the monthly salary for *anganwadi* workers, the payment for enumeration conveyed the importance and demanding nature of the work.

In contrast, DEOs in Bengaluru participated in a two-hour training to ensure that they could use the tablet PC. The training was focused on how to operate the tablet for data entry and retrieval. Their training did not include a discussion of the larger importance of the survey, how to conduct an interview, or the meaning of SEC survey questions, as it was the enumerator who was expected to conduct the interview. The regional coordinator for data entry operations at the start of the survey describes the DEOs' subordinate position:

> If a house is clearly made of mud, but the respondent says stone, the enumerator will tell the DEO to enter mud and the DEO must do so. While the enumerator has the authority to cross-examine, the DEO does not. But rarely does the enumerator cross-examine because they want to get the work done.[19]

The regional coordinator was clear: if there was an obvious discrepancy between what the respondent answered and what the data collectors observed, the enumerator had the responsibility to intervene—although they rarely chose to do so. The DEO was expected to enter in the tablet whatever the enumerator told them. Similarly, every local and state administrator that I spoke to put forth the view that enumerators would take the lead during the interview and direct DEOs as needed.

The envisioned role of an enumerator was much more limited in practice. The data collection teams that I observed fell into one of three categories: the DEO became the de facto enumerator (as in the opening vignette); the enumerator conducted the interview, but electronic data entry did not occur simultaneously; or the same individual served as both the enumerator and the DEO. Most data collection teams, like Mohan and Vijaya, used the resources at their disposal to conduct interviews in the most efficient way possible given the structure of the data entry program and the number of interviews they needed to complete.

In the first instance, the tablet and preloaded data from the NPR provided the DEO with tools that they could deploy to build trust with the household respondent. During the opening seconds of the interview, the tablet and preloaded data conveyed to the respondent that the data collection team was official and completing ongoing government work. The DEO could quickly build rapport and obtain buy-in from the respondent. This was especially important in locations like Bengaluru with negligible publicity in advance of the survey because urban residents did not have prior experience with the BPL survey.

This pattern emerged from the decisions to use tablets for real-time electronic data entry and then design a software program with limited flexibility; both decisions had the unintended effect of dramatically increasing the importance of the DEO. When the centralized planners provided specifications for the data entry program, they chose not to give DEOs flexibility to move between questions, record answers out of order, or submit the data for multiple questions at once. They took measures to limit the discretion of DEOs (for example, prevent the easy attribution of the head of household's religion or caste identity to other members of the family), but in practice the rigidity of the data entry program set the pace of the interview and made it easier for DEOs to ask questions instead of enumerators. DEOs generally started the interview to verify the preloaded NPR data (when the same household still lived in the home). Once the DEO started the interview, it was natural for them to continue as the main interviewer. As described in the opening vignette, it was most efficient for Mohan to ask questions and enter the responses based on the prompts in the data entry program, as the program did not allow for easy movement between non-consecutive questions. In this scenario, the role of the enumerator became limited to non-interview tasks. The demotion of the enumerator was neither planned for by local project managers nor reflected in the training materials.

I will note here that Vijaya and Mohan's pairing as a data collection team was not uncommon. The firm that oversaw local data entry operations realized that many enumerators, especially women, preferred a partner that they knew. As care work is gendered, most *anganwadi* workers are women, and they formed the largest single cohort of enumerators. Given the long hours, and intense nature of the job, women had an easier time signing up if they had the option to partner with a known individual, such as a family member or friend, or an unknown woman.[20] In contrast, centralized planners saw the two roles as distinct and coming from disconnected populations of government workers and technologically proficient DEOs. The consequences of this decision (that is, to encourage a known partner) are worth noting. In instances where a husband or another male family member became the DEO partner for their wife or another female family member, the gender dynamics of the relationship often shaped decision-making by the data collection team. The enumerator was more likely to be sidelined. Without prior training on the questionnaire, the DEO's personal understandings of caste, religion, and family (which varied considerably based on the background of the DEO) structured the production of data.

In a second set of cases, the enumerator was the lead interviewer. They asked questions, recorded the answers in writing, and then passed on the handwritten records to the DEO for entry into the tablet. In this arrangement, the enumerator played their envisioned role, but electronic data entry did not occur during the interview. I observed this type of "delayed" electronic data entry across three types of cases. In each case because the enumerator wrote the respondent's answers on paper, the respondent could see the recorded data.[21]

In one such case, the enumerator and the DEO stayed together while interviewing households outside of single-room dwellings in an urban slum. The enumerator wrote the answers on a sheet of paper to complete interviews quickly—as it was faster to complete an interview while hand-recording responses compared to electronically entering the data into the tablet. After completing several interviews, the enumerator would give the DEO her interview notes, which created an intermediary paper record of each interview. While the DEO entered the data into the tablet, the enumerator continued with additional interviews. In this instance, the data entry team did not take advantage of the NPR data during the interview. The neighborhood was adjacent to a building site and consisted of construction workers and their families who left and arrived with the ebb and flow of available construction work in the city and agricultural work in their home villages.

In a second instance, the enumerator and the DEO went door to door together in a lower-middle class neighborhood. The enumerator asked questions and the DEO recorded answers in her notebook. Each evening, after returning home, the DEO (often with the help of her teenage daughter) entered the handwritten data from her notebook into the tablet. This arrangement minimized their time in the field conducting interviews and allowed both women to take on the extra paid work but also permitted them to return home as quickly as possible for child and elder care.

In a third case, the enumerator went door to door in a government housing complex conducting interviews and recording responses on a sheet of paper, while the DEO remained in one location within the housing complex. The enumerator would periodically visit the DEO and drop off the handwritten data from completed interviews. The enumerator served as the lead interviewer and the DEO's role was limited to data entry. These three unimagined configurations that involved (intermediary) paper records elevated the position of the enumerator and aligned more closely with the interviewing role imagined by the BPL administrators.

A final scenario emerged from several on-the-ground realities in which the enumerator also served as the DEO. The difficulty of hiring DEOs (who were paid far less), the challenges of coordinating with another individual to conduct interviews (especially if both or one member of the team worked outside the home or had family responsibilities), and the technological literacy of some enumerators led to instances when the enumerator also completed data entry. The enumerator-cum-DEO would simultaneously conduct the interviews and enter responses into the tablet. In some instances, I observed the enumerator-cum-DEO entering the data real time, while in other situations they would record answers on paper and then later enter the data electronically—sometimes with the assistance of a family member. This configuration for the data collection team emerged as the government completed the training of enumerators in late 2011 and early 2012, but the private companies responsible for hiring and training DEOs struggled to recruit a sufficient number of DEOs. I observed the regional coordinator in charge of data entry attend the end of enumerator trainings and announce to the newly trained enumerators that if they had family members or friends who wanted to serve as DEOs, or if the enumerators themselves wanted to train to be a DEO, either option was possible.

The elevated role of DEOs described in the first scenario emerged from a mismatch between what centralized planners imagined and sought to protect against, and the required on-the-ground infrastructural work to build consent during actual household interviews. Enumerators—the government employees hired, trained, and paid to conduct interviews—often became sidelined when DEOs entered data into the tablet in real time. While the imagined technical role of the DEO required minimal training on the content and purpose of the survey, the structure and rigidity of the data entry program and the relative ease in consent-building when using a tablet led DEOs to become the lead interviewer in many instances. As I will describe in detail next, these realities created a situation where the enumerator training (however limited for the caste-related questions) became even less impactful in the production of data. Instead, the DEOs' perspectives on caste (as well as on religion and gender) took on a prominent role in the production of social data on many occasions. In addition, in areas where the training of enumerators remained silent about caste and religion, and the enumerators conducted interviews (as described in the second and third scenarios), their understandings of caste and religion structured the production of data.

Enumerating Caste and Religion

During enumeration, data collectors bring "to life" question-and-answer options from a survey.[22] Processes of enumeration and classification create "social equivalents" among individuals and groups that reflect a historically specific understanding of a social identity.[23] Interviewer understandings, respondents' points of view, and prevailing social norms also play a role in placing individuals or households in particular answer options for gender, caste, religion, or other social identities during the household interview. For example, in the opening vignette, Mohan's nonverbal cues help re-classify and simplify the respondent's response to the question about caste. Certain information is discarded during this process. In this instance, the data collector's own worldviews and understanding shaped the creation of social equivalence for all members of the family who are categorized as "Brahman" instead of a localized caste identity. Data collectors may use their discretion (consciously or unconsciously) to classify individuals in a manner that varies with the expectations of centralized planners.

Processes of enumeration are also structured by earlier decisions. For a national census or survey, centralized administrators (often with input from external experts and the public) design a survey and imagine a set of accompanying practices to create "social equivalents" that often reinforce state administrative categories, make some groups visible, and keep others invisible. Chapter 4 describes key decisions by the executive bureaucracy, including the choice to implement real-time electronic data entry, create a two-person data collection team, decentralize publicity, design an enumerator training that did not discuss how caste operates for religious minorities, and only require DEOs to partake in a technical training. The previous section of this chapter connects how these earlier decisions shaped household interviews in urban Karnataka. With regard to enumerating caste, the SEC survey included a three-question sequence to capture individual-level data on religion and caste, as discussed in Chapter 4. The questions followed the general structure of Census 2011 with three changes: the religion question was open-ended, with no categories provided; the first (of two) caste questions included the answer options of "no caste/tribe" and "other," along with "SC" and "ST"; and for the second caste question, enumerators recorded a specific caste for everyone except those who responded "no caste" to the first caste-related question (see Table 5.1).[24] By not including the answer options of "OBC" and "general" in the first caste question and limiting the use of caste lists to SCs and STs in the second caste question, survey administrators planned for the bulk of commensuration to occur after the collection of caste-wise data. For individuals classified as "other" for the first caste question (that is, up to 77.5 percent of the population), analysts would have to correct discrepancies in the spelling of caste names and decide whether to publish a particular subcaste or group several subcastes together under a broader category when "cleaning up" the responses to the second caste question. While executive bureaucrats planned for commensuration to occur during the finalization and tabulation of the caste-wise data, their earlier decisions regarding question-and-answer options and ground operations for the survey also affected processes of commensuration.

Based on my observations of household interviews in Bengaluru between December 2011 and February 2013, I trace the widening gap between the demands made by caste census advocates and the data that emerged (or failed to emerge) through household interviews during the SEC survey ground operations. I find that the collected data on religion and caste often conceptualized these identities at the household level instead of as individual-level identities. In addition, the "no caste" answer option was generally operationalized to hide the caste identities of caste-privileged groups. Finally, the data collected for many religious minorities provided unreliable information about caste.

Table 5.1 Religion and Caste Questions and Answer Options in SEC Survey (2011)

Religion (12)	**Caste/Tribe Status (13)**	**(14)**
(write name of religion in full)	Give code: Scheduled Caste (SC)-1 Scheduled Tribe (ST)-2 Other-3 No Caste/Tribe-4	If code 1, 2 or 3 in Column 13, write names of caste/tribe. If code 4 in column 13, put 'X'

Reproducing Caste and Patriarchy through Household-level Caste Data

Advocates for a caste census wanted individual-level data on caste and religion. The planners of the SEC survey were not naïve to the discrepancy between the data on caste and religion that they sought to capture, and the common perception and experience of caste and religion as family identities. During in-person enumerator trainings and in written training materials, the instructions to enumerators conveyed that religion and caste data should be gathered from each member of the household. An excerpt from my field notes from the second day of an enumerator training in central Bengaluru illustrates this point by the trainer—who oversaw the instruction of approximately forty enumerators in this location:

> "When you ask them about their religion, record whatever they say. If they say Hindu, Christian or Muslim—whatever they say record it. If they say, 'I don't have a religion, I am atheist'—record that. Write what they say in full. Ask this question about every member of the household. It is not necessary for every member to have the same religion, okay? Don't assume that just because the head of household is Hindu that the other people in the house are. You must ask, okay?" The lead trainer looks around the room as he is talking to make sure that everyone understands his point.[25]

The trainer emphasizes that enumerators should ask about the religion of *each* member of the household and not assume a shared religious identity. Like the message in the enumerator training, the data entry program required DEOs to finish recording individual-level information for one member of the household before inputting answers for another member. The program's structure prevented DEOs from consecutively entering religion and caste data for every member of the household. However, enumerators' and DEOs' beliefs and worldviews, as well as local social norms that structured the collection of data, led data collectors to often assume otherwise.

My observations suggest that the collected responses on caste and religion from Bengaluru are better conceptualized as household-level data. From 350 observations of survey interviews, I only observed one interview where the data collectors asked about religion or caste of more than just the head of household. While interviewing a lower-middle-class family, which immediately appeared nontraditional in multiple regards, I witnessed the enumerator asking about the caste of the wife after she had already collected information about the caste of the husband. The couple was young (in their early twenties) and living on their own in an unofficial rooftop apartment with the husband's younger sister. The arrangement suggested that this could be a "love marriage" and opened up the possibility that they were of different caste and/or religious backgrounds. Also noteworthy is that the enumerator played an active role in asking questions during interviews.[26] Although the enumerator and the DEO did not know each other prior to collaborating on the SEC survey, the enumerator was senior to the DEO in age and position, and both were female. Both members of the data collection team were also significantly older than the three young adults that comprised the household. While the enumerator was a teacher in her forties, the respondent was in her early twenties. These differences made the enumerator's inquiry about the respondent's caste (after the respondent shared information about her husband's caste) seem less intrusive and disrespectful. Almost universally, I observed interviewers record the caste and religion information shared about the head of household for all other individuals within the family.

Why did enumerators and DEOs—who asked about the sex, age, and occupation of each member of the household—so readily attribute the caste and religion of the head of household to remaining family members? Religion and caste are widely perceived as identities that one is born into and not individual choice, even in twenty-first-century Bengaluru. Data collectors most commonly conceived of religion and caste as family identities like most Indians.[27] They regularly classified all individuals within a household as "social equivalents" for caste and religion, despite repeated messages to the contrary in enumerator trainings and a data entry program that prevented consecutively entering caste and religion data for all household members.[28] After discussing that a head of household was "Hindu" or "Muslim," it made little sense to most data collectors and respondents to talk about the religion of other people in the household. From the perspective of prevailing social norms in a country where only 5 percent of marriages are inter-caste, if an interviewer asked about the religion and caste of each member of the household, it was repetitive at best and might disrupt the process of building rapport by suggesting that individuals living within the same household could belong to different religions or castes.[29] In addition, people usually do not discuss exceptions to the norm with strangers. As a result, this pattern of commensuration—attributing the head of household's caste and religion to other members of the family—was nearly universal in my observations.

While social norms structured how data collectors asked questions about caste and religion, the data collection technology reflected the views and predispositions of the centralized administrators who wanted individual-level data on caste. As such, the data collection teams created ways to circumvent the technology by either remembering the caste and religion of the head of household and entering the same data for subsequent members of the household or recording answers on paper during the interview and later entering the handwritten notes into the tablet.

While this observed practice of commensuration aligned with prevailing social norms, the same social norms helped to reinforce Brahmanical and patriarchal power relations. The enumerator training communicated that interviewers should not assume that the head of household is an adult male. Yet I observed data collectors and respondents consistently identifying adult males as the head of household.[30] The practice of attributing the religion and caste information of the (male) head of household to all other members of the family reinforced patriarchal family norms. While most households are made up of individuals who share a caste identity due to the reproduction of caste through caste endogamous marriage, important exceptions exist. In situations where individuals have had inter-religious or inter-caste marriages (or have changed their religion), the data incorrectly reflected that all household members shared the identity of the head of household. In these instances, processes of commensuration masked changing patterns of family formation. This pattern of classification reinforced power structures and perpetuated traditional social relations (that is, endogamous marriage practices that reproduce caste) even if households themselves are potentially challenging Brahmanism and patriarchy.

The Construction of "No Caste" and Obscuring Caste Privilege

In Bengaluru, respondents were aware of neither the project nor the questions and answer choices due to the lack of publicity around the SEC survey. For the (first) categorical question on caste, respondents could not have anticipated the option of "no caste" because it had neither existed in recent censuses nor on government forms in independent India; the response of "no caste" was not known to be an option. The designers of the SEC survey were aware that for the average Indian, such as a government enumerator, the "no caste" option did not make immediate sense.

During the first day of an enumerator training in Bengaluru, my field notes document how the trainer, who was a university professor, emphasizes that respondents do not have to provide information about caste:

> "Now, and this is important, if someone says they have no caste or tribe then select option '4: no caste/tribe.' Do not question them or persist that they should give you an answer. Especially in urban areas like Bangalore there will be people who say, 'no caste' and you should record their answer without troubling them." The trainer pauses for a second and looks around the room making certain that the trainees understand his point. Then he continues, "For question 17, write down their specific caste name, if you selected 1, 2, or 3 for the prior question. Whatever caste they say, you write it. Remember, this is the caste/tribe census."[31]

To ensure that enumerators and their supervisors conceptualized "no caste" as a real option, the trainer reiterated this point during a role-play exercise on the second day of the training. This excerpt from my field notes picks up midway through the role play, in which the trainer pretends to be a household respondent and Srinivas (who would later supervise six enumerators) acts as an enumerator:

> Srinivas looks up from the paper questionnaire and asks the trainer, "What is your religion?" The trainer replies, "I am Hindu." Srinivas nods and records the answer on his paper questionnaire. He then looks up again and asks, "What is your caste?" The trainer proudly responds, "I have no caste." Srinivas pauses and looks at the trainer, trying to determine whether he should probe further. He can't seem to decide if this is a situation that requires greater scrutiny or if he should just record the trainer's response. Around the room, the other trainees carefully watch the interaction unfold. Srinivas then breaks out of character to ask, "Sir, how should I proceed?" The trainer breaks out of his role, looking first at Srinivas and then the rest of the room. He explains, "In a metro city like Bangalore, there are going to be people who don't want to give any information about caste. You should not bother them or push them to give an answer. Simply mark 'no caste' for question 16 and then record an 'X' for question 17, where you are asked to enter a caste name."[32]

The role play continued and the trainer further illustrates the point when Srinivas seeks to collect information about the second household member:

> Srinivas asks, "And, your wife's religion, sir?" The trainer replies, "She is also Hindu." Srinivas nods and records the answer. Then he asks, "What is her caste?" There is a silence across the room and several enumerators lean forward in anticipation of the trainer's response. He states, "She does not have a caste." Srinivas nods, with a look of understanding on his face; he now knows that he should simply record "no caste" and move on. Elsewhere in the room there are a series of half-laughs, grins and head shaking from side to side. A few enumerators exchange a glance with a friend or acquaintance sitting nearby.[33]

The trainer's response of "no caste" challenged the prevailing sensibility for most government workers who had knowledge about the caste of their colleagues, neighbors, and other people whom they regularly encountered. As such, the trainer chose to emphasize that "no caste" was an acceptable option. In his portrayal, the answer naturally emerged from the respondent when answering the question, "What is your caste?" The enumerator did not offer up the answer option of "no caste."

Embedded in the training was also a message about the type of respondent who might respond "no caste." The trainer conveyed that the "no caste" option might be relevant among a subset of educated and well-to-do urban households. While the government attempted to introduce the concept of "no caste" to enumerators through the training, negligible publicity and media coverage meant that households themselves did not know about the SEC survey or the "no caste" answer option.

My field notes from the enumeration of a single-family home in central Bengaluru in March of 2012 provide an example of how the "no caste" option emerges relationally in the back and forth between data collectors and respondents.

> Before we knock on the front door of a large single-family home, Vijaya [the enumerator] puts a SEC Census sticker outside the door. A middle-aged man, who turns out to be the son of the head of household, arrives home as we are standing there. He looks at the sticker and reads aloud, "Socio-Economic and Caste Census." He then says, "Caste census? Why? We are all one caste—we are all Indians." Mohan, the data entry operator, quickly replies, "No problem, sir. That is also an option." Ten minutes later, after we are seated inside and midway through the interview, Mohan gets to the religion and caste questions for the head of household and offers up to the respondent, "I'll put 'Indian' and then 'no caste'." Everyone gathered in the hall nods in agreement. The enumerator says to Mohan, "You can also put 'Hindu' and then 'no caste'." Mohan replies as he continues to enter data into the tablet, "Hindu, Muslim, Christians—all are Indian." For each member of the household, Mohan enters "Indian" for the religion field and chooses the "no caste" option for the categorical question on caste.[34]

Mohan sensed that asking about caste would offend the sensibilities of the respondent given his statement about the SEC survey prior to the start of the interview. The respondent had a high social class—which was immediately obvious given the size of the house and its prime location—especially compared to Mohan, who was Brahman, yet ran a small copy shop and lived in a modest two-room rental with his wife and son in a congested part of the city. Mohan offered up the "no caste" option without even posing an open-ended question—as he usually did. In my observations of Mohan and Vijaya, and other data collection teams, this occurred most commonly among a subset of well-to-do households who identified as "Hindu" but who did not see themselves as beholden to a caste identity. The data collectors would ask a question and offer the option of "no caste" within the same breath: "Sir, your caste? If you wish you can say 'no caste.'" Once offered, the option was almost always taken and there was a fairly clear signal beforehand that it would be.

In contrast, no one hesitated to talk about caste and the interviewers never offered up the "no caste" answer option during my observations of interviews in an urban slum. Again, an excerpt from my field notes:

> When I arrive in the slum, Shalini [the enumerator] and Sumitra [the DEO] are seated one lane/row over from where we left off the previous night. They are sitting on a stone with four to five children around them. They complete the details for another three families by asking questions to the kids. One young boy—probably 10—answers the question about his family, which consists of him and his two grandparents. He knows their names and Shalini estimates their ages, based on the child's feedback. When Shalini asks the name of the boy's parents, the other kids shout, "He doesn't have any" [because the boy doesn't live in the slum with his parents] but he proudly says, "Baiappa and Baiamma." When Shalini asks him about his caste, the boy pauses and then begins to reply, but before he can answer an older boy standing next to him quickly responds, "He is SC—Chalvadi." This is an easy question for anyone living in the slum because of the clustering by caste and religion in each lane.[35]

Unlike the first example from a wealthy neighborhood where almost all the residents were from privileged castes, more than 80 percent of households living in this slum for construction workers were Dalits. They had been providing caste data for each decennial census and never conceived that "no caste" was an option in a government survey. The lack of local publicity surrounding the introduction of the "no caste" answer meant that the average person did not know that "no caste" was a possible response. While the government attempted to introduce the concept of "no caste" to enumerators, it did not make a similar effort to ensure that households knew about the "no caste" option. At the same time, this entire community was likely eligible for BPL status and by disclosing SC identities they relayed information relevant for BPL identification. In this community, where there was almost complete residential segregation by caste such that all families living in the same lane shared a caste identity, individuals had considerable knowledge about their neighbors. Respondents repeatedly explained, "This whole lane is Chalvadi"—almost irritated when enumerators asked the question for each household. Enumerators never offered up the "no caste" answer nor did respondents hesitate to answer the question. I observed a similar pattern in another slum-like colony in another peripheral location in the city.

Data collectors and household respondents jointly constructed the "no caste" answer option during the household interview. In the first example, the DEO sought to smooth things over when he offered to record "Indian" and "no caste" after the respondent signaled that he did not agree with the collection of caste-wise data. In other similar cases that I observed, respondents did not express an explicit disdain for providing caste information. However, the high social position and visible wealth of the family led data collectors to assume that the respondent might not want to provide caste details and the interviewers conveyed the "no caste" option. Given the class distance between the data collectors and households in these cases, data collectors were particularly sensitive to avoid offending the sensibilities of respondents, limit awkward or embarrassing moments, and increase the likelihood for participation in the interview without tension. The potential for embarrassment caused by asking about caste correlated with class and caste. In the urban slum, everyone talked about caste, and it openly structured where people lived. No one hesitated to ask about caste and the "no caste" option was never offered up in my observations. For these residents, the possibility of accessing a seemingly universal position, such as "no caste" in a government survey seemed disconnected from daily life.

The deployment of "no caste" among elite Indians in this case varies in important ways from social movements that have organized to eradicate caste hierarchy. Many anti-caste leaders and organizers, beginning in the late nineteenth and early twentieth centuries, intentionally disavowed their caste identities and promoted inter-caste marriage in their efforts to dismantle the caste system. Yet, in this case, the option of "no caste" emerged within a section of the social and economic elite who perceived that caste no longer shaped their life opportunities; these households commonly resided in well-known caste-elite enclaves with cleaner streets and comparatively reliable services and amenities. The construction of the "no caste" answer institutionalized castelessness in the caste-wise data. Caste-oppressed individuals remained hyper-visible while caste elites had the option to conceal their caste, which invisibilized caste-based power. Along with these instances, interviewers deployed the "no caste" answer option for religious minorities in some instances, which I discuss later in the next section.

The Failed Enumeration of Caste for Religious Minorities

During re-enumeration exercises in a densely packed predominantly Muslim neighborhood in the periphery of Bengaluru in early 2013, I witnessed a recurring pattern.[36] The following excerpt from my field notes describes one such interview:

> A woman opens the front door of a two-room home. She nods at the three of us, as Patma [the enumerator] asks her, "Name?" The woman answers, "Ansari. Faizal Ansari"—giving the name of her husband. Patma continues, "Your name?" The woman gives her name and then slowly lists the names of her children. Shareen [the DEO] records the woman's responses in a school-sized notebook, using a pencil. As the respondent states each name, Patma asks about the person's birthdate. The woman doesn't know her husband's date of birth or her own—so Patma asks their age. After recording the birthdates of their four children, Shareen asks about the names of the respondent's in-laws, how much schooling each person has completed, and the type of work they do. The woman responds and Shareen records the information next to each name in her notebook.
>
> The interview continues and Shareen asks, "Religion?" The woman responds, "Muslim." Shareen follows up, "Caste?" The woman responds, "Ansari." Shareen shakes her head, "Sunni or Shiite?" The woman looks at her for a moment. Shareen offers up, "Sunni, no?" The woman nods. Shareen records "Sunni" and continues with the interview. Shareen asks, "Do you have mobile? A two-wheeler or a four-wheeler? Washing machine?" After the woman explains that she only has a mobile, Shareen thanks her and we begin to walk away. As we are leaving, the woman asks us about the purpose of the interview. Patma simply replies, "The census."[37]

In this interview, the data collectors (whom I observed for two weeks) and their supervisor perceived two caste divisions within Islam: Sunni and Shiite. They systematically recorded "Sunni" as the caste for all the Muslim families in this neighborhood—even in instances when the families offered up a caste identity. Sometimes households offered up "Sunni" when asked about their caste but not always, as was the case with the Ansari household.

The interviewed families knew the data were being collected for the government, but at the time of data collection the larger implications were not understood. Caste information might be necessary to identify Muslims eligible for affirmative action benefits. For example, "Ansari" falls in the administrative category of "OBC" in many regions of the country while in Karnataka the

OBC reservation quota includes a 4 percent subcategory for Muslims. In the vignette, the DEO collects data on sect (which may be an important identity for many families), instead of the information that identifies Muslims who are part of historically oppressed castes.

In my observations, I saw a variety of patterns unfold across data collection teams. In other parts of the city, where I observed the enumeration of Muslim families in majority-Hindu neighborhoods (usually in upper-middle-class areas), the data collectors recorded Muslim or Islam as the religious identity and entered the conveyed caste identity (for example, Sherriff, Pathan, etc.). In contrast, in a lower-middle-class Muslim neighborhood in the heart of the city, I spoke to an enumerator at the local charge center. I asked if he had run into any problems when collecting caste-wise data. He explained that since they were collecting data in an entirely Muslim neighborhood, and Muslims "do not have a caste," he never asked about caste. They simply recorded "no caste" for everyone.[38] As such, households in this area did not even have the option to answer the caste question. The worldviews of data collectors heavily shaped processes of classification and the creation of social equivalents with regard to caste for Muslims.

I also observed the enumeration of Christian families in predominantly Hindu neighborhoods. In my observations of twenty to twenty-five Christian households, across multiple neighborhoods, I saw data collectors record "Christian" for the question on religion and denomination (for example, Methodist or Roman Catholic) for the question on caste. These answers emerged during the back and forth between the data collectors and respondents. Data collectors usually simply asked, "Religion?"—or based on visible images of Christ or a crucifix asked, "Are you Christian?" In the former case, the respondent usually replied "Christian" or with a specific denomination, such as "R.C." (that is, Roman Catholic). The data collector then recorded "Christian" or "Christianity" for religion and "R.C." or another Christian denomination for caste. In the second scenario, when the data collector asked, "Are you Christian?" and the respondent would agree, the interviewer would follow up by asking, "What type?" The respondent usually provided information on denomination, which the data collector then recorded as the caste. While denominational identities are no doubt extremely important to households, they failed to capture the information necessary to identify Dalit Christians who are eligible for affirmative action in the OBC administrative category in certain states.

Unlike the "no caste" answer option and the attribution of the head of household's caste and religious identities to other household members, the enumerator training was largely silent on how religion and caste intersect, and what caste meant for religious minorities in India. Beyond the discussion of the relationship between religion and caste for SCs and STs (which was standard material from decennial census trainings), the enumerator training manual and in-person program did not provide information on how caste relates to Muslims and Christians. Data collectors' and households' varying perceptions of what caste meant for Muslims and Christians filled the gap and heavily shaped the production of caste data for religious minorities in India.

Most Muslims and Christians in India are converts (or descendants of converts) who hoped to flee the discrimination they experienced within the caste hierarchy. But since caste structures so many aspects of life, it has not been easy to leave caste behind. While Dalit Muslims and Dalit Christians are not recognized as SC by the government, like their Hindu, Sikh, and Buddhist counterparts, they continue to face caste-based discrimination even after converting.[39] They often worship in separate churches or religious spaces and face discrimination within their own religious community, and from the larger society, because of their caste and religion. The caste system has reproduced itself within minority religions in India.

The glaring absence of a plan for how to enumerate caste for over 200 million Muslims and Christians aligned with the ideology of Hindutva. The invisibility of Muslims and Christians in the design of the caste questions (and answers) and the lack of attention to how these groups should be classified by caste in training materials spoke to their status as "otherized" groups.

Looking Back and Ahead

Contextualizing this chapter within the broader narrative, I document several interconnected steps of bureaucratic deflection that flowed from earlier decisions made by senior bureaucrats. First, the lack of publicity at the start of data collection left the burden of consent-building to survey interviewers. Households had limited knowledge about the SEC survey and caste-wise enumeration, so interviewers had to convince them to participate. Data collection teams often built consent for the interview by alluding to "the census" (such that their work seemed connected to the recently completed decennial census), showed their tablets to signal official government work, and used the NPR data to indicate prior knowledge about a household. Second, and related to processes of consent building, when interview teams recorded data real time in a tablet, the role of DEOs became elevated and DEO perspectives structured the production of data. Since most DEOs only participated in a short technical training on how to use the tablet and data-entry software, they lacked a sociological understanding of the survey questions and answers. Third, the earlier decision to use the decennial census as the basis for the caste-related questions and the enumerator training materials led to four observable phenomena: the caste questions in practice became "open-ended" for most individuals, as the "OBC" and "general" category answer options were not included in the first question on caste and the second question lacked caste lists to select from for individuals who were not SC or ST; the open-ended answer options, coupled with insufficient training on the caste-related questions, led to the lack of a common metric to commensurate caste within and across the OBC and "general" categories; the failure to educate data collection teams about how caste shapes the lives of religious minorities created fundamental confusion over the caste-wise enumeration of Muslims and Christians, including the difference between caste and sect in Islam and between caste and denomination in Christianity; and the inclusion of the "no caste" answer option created an opt-out for caste elites. These decisions protected privileged castes, reinforced patriarchal norms, and limited data collection on caste from Muslims and Christians, in line with ideologies of Brahmanism, castelessness, and Hindutva.

Three related patterns that I observed during the ground operations of the SEC survey in Bengaluru shaped the construction of "social equivalents" for caste and religion data and illustrate the limitations of the collected data. First, I consistently observed the attribution of the (male) head of household's caste and religious identity to other members of the family. This practice reinforced existing hierarchies of power by gender, religion, and caste while supposedly producing "apolitical" and "neutral" data. While SEC survey data on religion and caste reflected prevailing social norms, they also perpetuated them where they no longer applied, such as in the case of inter-caste or inter-religious marriages. Second, interviewers made the "no caste" answer option available in certain instances when they observed class and caste privilege—such that asking about the caste or religion of an individual might offend the respondent and make them less willing to participate. This practice reinforced an ideology of castelessness by offering the "no caste" option to a limited set of urban elites. The option to "opt out" was absent for the majority and selectively applied by data collectors and protected the invisibility of caste-based power.

I also observed data collectors record "no caste" for Muslim households in a lower-middle-class neighborhood, without asking respondents about their caste. The failure to ask about caste in this Muslim enclave forwent the opportunity to document internal differentiation among Muslims, and intersecting experiences of religious, caste, and class-based discrimination, as the Pasmanda movement has highlighted.[40] This practice reinforced a Hindu nationalist ideology, which is inattentive to structural differences in life experiences and opportunities while constructing Muslims as "the other." A third and related pattern that I observed involved the failure to enumerate the caste of religious minorities and instead interviewers enumerated sect or denomination. This gap must be contextualized within a longer history of Muslims and Dalit Christians struggling to gain recognition as equal citizens, as they belong to religions that are seen as "non-indigenous" and "foreign." State policy has excluded Dalit Muslims and Christians from access to SC reservation benefits. The production of caste data during the SEC survey shows the continued struggle to recognize religious minorities, and the state's disregard for producing reliable caste-wise data for the implementation of redistributive and anti-caste policies. While I observed these simplifications in the metro area of Bengaluru, similar patterns are likely to unfold in places with comparable social norms and histories of caste politics.[41]

My observations of processes of enumeration and classification trace how fraught commensuration can be in a state project relegated to the periphery. Attention to the social and political embeddedness of processes of enumeration and classification is essential, especially if the motivation for collecting survey data is to aid in the dismantling of discriminatory systems related to the distribution of power. Given the pervasiveness of Brahmanism across key institutions of society, the caste survey was particularly susceptible to being derailed.[42] Following the nationwide collection of caste data, a tumultuous process to bury the collected caste-wise data unfolded, as the next chapter details.

Notes

1. Interviews with this data collection team took place in a central neighborhood in the city and were in a spoken mix of Kannada and English. If the household did not speak Kannada, interviews were generally conducted in English or Hindi. The data entry program—which became the de facto questionnaire for this interview team—was in English. Like all the interview teams that I accompanied, the members of this data collection team were fluent in at least three languages.

2. This vignette is from my February 27, 2012, field notes. I use pseudonyms for individuals in my field notes.
3. In 2006, the city government of Bangalore changed its name (back) to Bengaluru, with approval from the state government of Karnataka later that year. In 2014, the union government officially approved the change.
4. One such example of a clue that data collectors observe to guess the religion of the household are *kolam/muggu/rangoli*—or decorative art drawn with rice flour and chalk powder—in the entryway of the home. However, most data collectors I observed asked about the religion of the head of household even if they might be able to deduce it from the household name and other visible clues.
5. Vijaya and Mohan are vegetarian and Brahmans from another region of Karnataka. When we entered Christian and Muslim households, particularly on Saturday or Sunday afternoons or evenings, families might be preparing meat. Vijaya often complained about the smell after an interview and sometimes rushed through an interview a bit if they found the smell of meat too strong for their preferences.
6. Anderson, *Imagined Communities*; Curtis, *The Politics of Population*; Desrosières, *The Politics of Large Numbers*; Guha, "The Politics of Identity and Enumeration in India c. 1600–1990"; Loveman, "The Modern State and the Primitive Accumulation of Symbolic Power"; Porter, *The Rise of Statistical Thinking, 1820–1900.*
7. For surveys and censuses that do not involve face-to-face enumeration, technologies that replace the enumerator also play a key role in processes of enumeration.
8. Sociologist Michael Rodríguez-Muñiz describes consent-building as the on-the-ground work of creating legitimacy for the census, mobilizing households to participate, and making marginalized groups legible, as discussed in Chapter 1. Rodríguez-Muñiz, "Cultivating Consent."
9. In contrast to enumerator trainings, which involved the translation of materials from the center to the state, each district in Karnataka was expected to generate publicity materials to advertise this project. Publicity was entrusted to the Principal Census Officers in each district; in the case of Bengaluru, the Commissioner of the BBMP (that is, the municipal government) was in charge of "the local requirement of publicity." Interview with government employee, August 7, 2012.
10. A short TV advertisement ran sporadically on a major Kannada TV station during the final four weeks of data collection (April to May 2012). This ad explained that an enumerator would visit each household and collect details about social and economic conditions. While this type of ad could have been very effective at the start of the survey to communicate the purpose of the survey and ensure that households participated, the timing of the ad was very late in the data collection process.

11. Newspaper articles, in the high-circulation Kannada press, were few and far between. The leading vernacular newspaper was quiet on this topic, in a state with active coverage of caste (for example, other caste-related topics published in articles in *Vijaya Karnataka* during this same period covered conferences, meetings, and statements of caste associations, efforts by political parties and leaders to organize along caste lines, and public policies targeting specific castes). The day after the state cabinet approved that the state-level equivalent of the Ministry of Rural Development (that is, the RDPR) could conduct the survey, and it would also be conducted in urban areas, a short news announcement was published on page 11 of the largest-circulation Kannada-language newspaper. "Urban Caste Census," *Vijaya Karnataka*, September 16, 2011.
12. Local editions of the English-language *The Times of India* (circulation of 517,000 in Bangalore), *Deccan Herald* (local circulation of 201,000), and *The Hindu* (local circulation of 150,000), which are the three leading English newspapers in Karnataka's major cities, had sporadic coverage of the caste census in the period leading up to the canvassing of the survey in Karnataka. A handful of articles mentioning the SEC survey discussed the costs of the exercise, the agencies that were involved, the use of PC tablets, that *anganwadi* workers would serve as enumerators, and the project's connection to previous BPL surveys. While some district-specific news articles discussed the SEC survey during the period of preparatory work, on the whole, Kannada and English newspaper coverage was limited with regard to communicating the purpose, scope, and content of the upcoming SEC survey. At the same time, a few regions published articles with particulars about the SEC survey. Relatively detailed English-language articles from Bidar, Kolar, Dakshina Kannada, and Hubli-Dharward described the upcoming SEC survey. These articles mentioned the timeline for data collection, the appointment and training of local enumerators, and how DEOs would use tablets (*The Hindu*, September 29, 2011; *Deccan Herald*, October 3, 2011; *Deccan Herald*, October 4, 2011; *The Hindu*, November 10, 2011). An article from Bidar described the project as a caste census and the enumeration of families living below the poverty line, while the article from Kolar described it as an effort to collect statistics "to study the economic and social condition in the country, so as to frame rural development projects" (*Deccan Herald*, October 3, 2011). Only articles from Mangalore and Hubli-Dharward discussed the historic nature of the caste-wise data collection; both mentioned that the last caste-wise enumeration was completed in 1931. Two brief articles from Chitradurga and Bangalore discussed the recruitment and preparation of the SEC survey staff (*The Hindu*, October 17, 2011; *Deccan Herald* November 10, 2011).
13. Interview with SEC survey enumerator, December 9, 2011.

14. During the primary period of data collection in Karnataka from December 2011–May 2012, media coverage remained minimal. During this period, *Vijaya Karnataka* had two articles related to the SEC survey. Both were published in December 2011 and discussed how state leaders reached out to members of their caste communities and advised how to answer the caste question during the SEC survey ("Enter as the Word 'Kunchatiga' Only," *Vijaya Karnataka*, December 19, 2011; "Enter as 'Ganig' Only," *Vijaya Karnataka*, December 26, 2011). *The Hindu* had sporadic coverage of the SEC survey and published about ten articles, which touched upon different aspects of the SEC survey operations. More than half of these articles were brief consisting of a short paragraph which mentioned when the survey would start in particular locations ("Census Process Begins Today," *The Hindu*, December 9, 2011; "SECC Gets Underway," *The Hindu*, January 13, 2012; "Caste Census 2011," *The Hindu*, January 27, 2012; "Caste Census Gets Under Way in Madikeri," *The Hindu*, January 28, 2012; "Caste Census from Friday," *The Hindu*, February 8, 2012; "Caste Census," *The Hindu*, February 11, 2012). Two other articles were similar to, and predated, the *Vijaya Karnataka* articles on political leaders trying to mobilize communities to provide specific subcaste information ("Weavers Want Uniform Name in Census," *The Hindu*, December 11, 2011; "Appeal," *The Hindu*, December 17, 2011). The most detailed article on data collection within the state was published in January 2012. Across the English and Kannada newspaper coverage, this article was singular in the details that it provided about the enumeration of caste and the discussion of the identification of BPL families (Vijesh Kamath, "Census Midway, Has Many Firsts," *The Deccan Herald*, January 22, 2011). This article touched upon important aspects of the survey, including the broader purpose of the caste-wise enumeration. The article also clarified that a socio-economic survey had previously been held in rural area every five years by the MoRD, but that this round of the survey had been extended to urban areas. The article discussed the optional nature of providing caste information:

> There are provisions to mention religion, caste and sub-caste, but it is not mandatory for a respondent to mention all, or any of these. Under the caste column, some options are given: Scheduled Caste, Scheduled Tribe and others. It is purely optional to respond to this column. (Ibid.)

The article quoted a senior officer for the SEC survey operations in Karnataka as saying that it is not mandatory for people to provide information on their subcaste. The article again reinforced this point when it stated, "In the section on caste and tribes, the questionnaire includes a 'no caste/tribe' option for those who do not wish to indicate their caste, or belong to religions without a caste system" (ibid.). This article was unique because it both mentioned details about the enumeration of caste and described the significance of the exercise, instead of simply discussing logistical or timing aspects of the survey.

15. For example, during my fieldwork in Bengaluru, I only learned about one caste-based organization that did organizing related to the SEC survey. "Weavers Want Uniform Caste Name in Census," *The Hindu*, December 11, 2011.
16. Field notes, December 3, 2011.
17. Interview with SEC survey enumerator, December 5, 2011.
18. Enumerators earned INR 450 for attending the two-day enumerator training; INR 12,000 for completing data collection in four blocks (at INR 3,000 per block); and INR 6,000 for forty days of work based on a daily stipend of INR 150.
19. Interview with Transvision Manager, December 8, 2011.
20. At the end of the enumerator training, when an enumerator asked if her younger brother could partner with her as a DEO, the local representative of the private firm in charge of DEOs encouraged her and other enumerators to ask family members or friends who had computer literacy to sign up for the DEO training.
21. The official definition of a slum varies but colloquially refers to a crowded urban community with limited, unreliable, and often unsafe access to basic services, such as water, sanitation, and electricity, and insecure tenure which makes harassment or eviction possible or in many cases routine.
22. Curtis, *The Politics of Population*, 4. See also Wendy Nelson Espeland and Mitchell L. Stevens, "Commensuration as a Social Process," *Annual Review of Sociology* 24 (1998): 313–43.
23. Curtis, *The Politics of Population*.
24. Similar to the decennial census, pre-set caste and tribe categories were provided for SCs and STs. For all others, no caste lists were provided and the data collectors were expected to record whatever answer the respondent provided. In the decennial census, enumerators only asked caste or tribe names from respondents who identified as SC or ST.
25. Field notes, December 11, 2011.
26. Over time, this team had developed techniques to cope with the slowness of the data entry program by using both members of the data collection team in creative ways.
27. The two questions on disability fall into a category of question types that were conceived as individual characteristics but often (but not always) asked at the household level (for example, "Is everyone in the household healthy?" or "Does anyone have a disability?"). My observations of household enumerations and interviews of enumerated households thus reveal that the questions on "individual particulars" actually fell into three different categories of questions (asked as individual-level questions, asked as household-level questions, or asked of the head of household and their information was applied to all other members).

28. DEOs also recorded answers to other individual-level and household questions without specifically posing a question. For example, in the opening vignette to this chapter, Mohan uses the response to the question about occupation to answer the wage question; since the head of household works at a bank, Mohan concludes that he is a salaried employee who receives a monthly paycheck. Similarly, household-level questions about the structure of the house and goods possessed by the family were often not asked when the DEO made an assessment or decision based on what they observed. For example, for a poor household living in a slum with limited electricity, the DEO would not ask if the household had a washing machine. There was a logical hierarchy; if a household lacked electricity, the data collectors would still ask if they had a mobile phone but might not ask if they had a computer or washing machine. Data collectors made inferences based on responses to other questions, objects within or outside the home, and a combination of these factors. While these practices diverged from how central planners imagined that the household interview would take place, they seemed reasonable assumptions based on the available data and spoke to the gap between the length of a questionnaire that planners believed was reasonable and the amount of time and details that data collectors felt comfortable posing to a respondent (especially when the answers appeared obvious).
29. Keera Allendorf and Roshan Pandian, "The Decline of Arranged Marriage? Marital Change and Continuity in India," *Population and Development Review* 42, no. 3 (2016): 435–64.
30. The designers of the SEC survey repeatedly explained that the head of household should not be assumed. The enumerator's manual says, "Note that a household may also be headed by a female member. Please ask from the household who is the head and record accordingly. On your own, you should not assume that the household is headed by a male member." Ministry of Housing and Urban Poverty Alleviation, "Instruction Manual for Enumerators, Socio-Economic and Caste Census 2011- Urban," 20.
31. Field notes, December 3, 2011.
32. Ibid.
33. Ibid.
34. Field notes, March 2, 2012.
35. Field notes, December 9, 2011.
36. In Bengaluru, a small number of enumeration blocks were identified for re-enumeration. In the selected blocks, the total population for the SEC survey was noticeably smaller than the count generated during the 2011 decennial census (and there was no obvious explanation for this difference, such as the clearance of a slum).
37. Field notes, January 15, 2013.
38. Field notes, December 11, 2011.
39. As discussed in Chapter 4, the scholarship on caste in Muslim and Christian communities documents experiences of discrimination, untouchability,

practices of caste endogamy, and segregation by residence, occupation, and place of worship. Some key texts include Imtiaz Ahmad, "The Ashraf-Ajlaf Dichotomy in Muslim Social Structure in India," *Indian Economic and Social History Review* 3, no. 3 (1966): 268–78; Ahmad, *Caste and Social Stratification among the Muslims*; Ahmad, *Caste and Social Stratification among the Muslims*; K. C. Alexander, "The Problem of Caste in the Christian Churches of Kerala," in *Caste among Non-Hindus in India*, ed. Harjinder Singh, 60–65 (Delhi: National Publishing House, 1977); J. Tharamangalam, "Caste among Christians in India," in *Caste: Its Twentieth Century Avatar*, ed. M. N. Srinivas, 263–91 (Delhi: Penguin Random House India, 1997); Ahmad, "A Different Jihad"; Seik Rahim Mondol, "Social Stratification, OBCs and Muslims," *Economic and Political Weekly* 38, no. 46 (November 15, 2003): 4892–97; Sayyed Zainuddin, "Islam, Social Stratification and Empowerment of Muslim OBCs," *Economic and Political Weekly* 38, no. 46 (November 15, 2003): 4898–901; Anwar Alam, "Democratisation of Indian Muslims: Some Reflections," *Economic and Political Weekly* 38, no. 46 (November 15, 2003): 4881–85; Satish Deshpande and Geetika Bapna, *Dalits in the Muslim and Christian Communities* (New Delhi: National Commission for Minorities, Government of India, 2008); Prashant K. Trivedi, Srinivas Goli, Fahimuddin, and Surinder Kumar, "Does Untouchability Exist among Muslims? Evidence from Uttar Pradesh," *Economic and Political Weekly* 51, no. 15 (April 9, 2016): 32–36.

40. The Pasmanda movement has highlighted severe caste and class-based inequalities within the Muslim community in India. See Khalid Anis Ansari, "India's Muslim Community under a Churn: 85% Backward Pasmandas Up Against 15% Ashrafs," *The Print*, May 13, 2019, sec. Opinion, https://theprint.in/opinion/indias-muslim-community-under-a-churn-85-backward-pasmandas-up-against-15-ashrafs/234599/, accessed April 1, 2024.
41. For example, in regions of Tamil Nadu with a long history of oppressed-caste social movements, the newspapers of local caste associations printed the full SEC survey questionnaire just prior to the start of the survey period, and there was greater newspaper coverage of the SEC survey. In these communities, data collectors would not be the sole source of information regarding the survey, as many households would already be familiar with its details. The construction of "no caste" might also follow a different pattern during household interviews because a wider cross-section of the population would be aware of the "no caste" option. The patterns that I observed might vary where caste politics and publicity around the SEC survey unfolded quite differently.
42. Chapters 3 and 4 show how this process unfolded under the direction of executive bureaucrats and project planners in Delhi as they decided where and how to collect the caste-wise data.

6

Disappeared Data

The Life and Death of Caste Data

On July 3, 2015, four years after the MoRD started data collection for the SEC survey in the state of Tripura, it co-announced the release of provisional data for rural India. Finance minister Arun Jaitley and minister of rural development Chaudhary Birendra Singh—both of the BJP-led government that had come to power in 2014—published a press statement that included a link to SEC survey summary data on the MoRD website. The MoRD had analyzed and made public the data that it required for BPL identification, and envisioned numerous additional possibilities for using the collected data. The press release stated that the MoRD planned "to use the [SEC survey] data in all its programmes," specifically citing "Housing for all, Education and Skills thrust, MGNREGA [Mahatma Gandhi National Rural Employment Guarantee Act], National Food Security Act, interventions for differently able[d], interventions for women-led households, and targeting of households/ individual entitlements on evidence of deprivation, etc." It also pointed to other intended uses of the SEC survey data by policy planners across levels of government and stated that combining the SEC survey data with the NPR would allow coordination across programs "to simultaneously address the multi-dimensionality of poverty by addressing the deprivation of households in education, skills, housing, employment, health, nutrition, water, sanitation, social and gender mobilization and entitlement" and tracking the progress of households over time. The concluding paragraph of the press release argued that the "[SEC survey] truly makes evidence based targeted household interventions for poverty reduction possible."[1] The two summary tables in the press release included details on the percentage of households automatically included on the BPL list based on five possible parameters for automatic inclusion, one of which was "whether any member of the household was from a primitive tribal group" (a question from the household section of the SEC survey questionnaire). This data came from a question that had been field-tested in the 2010 pilot survey—so it was not an addition from when the caste-wise enumeration was combined with the BPL survey. The tables also included the percentage of SC and ST households, which previous BPL surveys had documented.

Notably, the press release was comparatively silent on the collected caste-wise data with only a brief reference to the "caste census":

> The districts and State Governments have carried out the [SEC survey] with the Ministry of Rural Development as the nodal Ministry. Ministry of Housing and Urban Poverty Alleviation carried out the survey in urban areas and the Registrar General, Census of India carried out the caste census. The provisional socio-economic data for Rural India has been released today.[2]

The MoRD's press release made no mention of the timeline for releasing the caste-wise data. The SEC survey and the "caste census" became two separate exercises in this explanation, with the ORGI carrying out the caste-wise enumeration. It was a puzzling statement given that a single survey instrument collected data for both the fourth round of the BPL survey and the "caste census."[3] At the same time, the statement highlighted the ORGI's role in designing the caste-related questions for the SEC survey and its receipt of the collected caste-wise data after the completion of the survey, while ignoring the ORGI's "success" in blocking the caste-wise enumeration from the actual census and therefore not having to publish the data as part of Census 2011. So where were the collected caste-wise data? What had happened to them?

After hearing about the release of the BPL data and silence surrounding the publication of the caste-wise data, I remembered an earlier conversation with a caste census activist in 2013. He predicted, "If the BJP win the 2014 elections, they will probably bury the data."[4] This chapter describes the final stages of compiling and verifying data from the SEC survey and traces what happened to the collected caste-wise data—which, as anticipated by the activist, the government eventually buried.

Finalizing the SEC Survey Data

Going from the completion of household interviews to the finalization of the collected data involved several planned steps. First, each interview team uploaded their data from the tablet to a server at their local charge center. Supervisors were expected to randomly check the data from each block.

If differences existed between the data gathered by the supervisor and the original data, supervisors were to conduct additional interviews to see if the discrepancies were isolated or if they signaled larger problems with the interview team's work. Once the supervisor approved the collected data, the interview teams created an abstract for each block. With the completion of the block abstracts, the enumerators and DEOs finished their work, and the draft data were uploaded to the NIC server. The compiled data then underwent a public verification process. A draft list of all household-level data (which excluded the caste-wise data) from each enumeration block was made public in the local ward or village office and other prominent public places.[5] Any individual could challenge the data by filing a complaint in the local *panchayat* (rural) or ward (urban) office. Individuals could request a change in their own household data or challenge data about another household. In addition, individuals could complete a form stating that their household had been omitted from the data collection process.[6] As the project unfolded, the managers decided to institute a detailed process for publishing the draft SEC survey lists at the district level and on the government's SEC survey website, to allow widespread access and verification. Local officials were expected to verify each claim or objection and hold a hearing within seven days. If an individual was not satisfied with the decision by the local officer, they had the right to appeal the decision at the district level.[7] Following the public verification period, states published finalized lists for each district, and the data could be tabulated, analyzed, and disseminated for policymaking purposes. A planning document explained that "at the end of the 31st day from the publishing of the draft report, the final list would be published" and sent to local government offices.[8]

Based on my fieldwork in Bengaluru, some differences existed between the built-in mechanisms for finalizing the data and the actual processes that unfolded on the ground. First, supervisors focused on overseeing data collection processes, but rarely double-checked the collected data in the field. Supervisors checked-in with data collection teams to monitor their progress, help troubleshoot any unexpected challenges, and ensure that data collection teams were completing their work. In general, most data collection teams took much longer than scheduled to complete the interviews because urban residents were often not home during the workday and, in general, had unpredictable schedules. Data collection teams adjusted to this reality in different ways: by shifting their work to late afternoons and early evenings, concentrating much of their interviews during the weekends, or returning to a missed household at a different time for a second visit. These adjustments increased their overall timeline for completing data collection. Residents of the city were generally unaware of the SEC survey, as discussed in Chapter 5, and some residents were therefore hesitant to share information with strangers for an unknown project. Data collection teams would either return and try to secure an interview with another family member, who might be willing to complete the interview, or report the household to the supervisor to see if the supervisor could secure an interview. By the time data collection teams finished their work, many were well behind their original schedules, so supervisors reviewed the collected data at the local charge center to identify omissions or inconsistences, and if they noticed any problems asked the data collection teams to address the issues.

In addition, some entire blocks required re-enumeration. In early 2013, I observed groups of blocks within a single ward and an entire ward being re-enumerated, although this process was not originally planned for by SEC survey administrators. New enumeration teams were sent to recollect fresh data in each block selected for re-enumeration. In Bengaluru, I was told by ground-level staff (enumerators and their supervisors) that this occurred when the first round of collected SEC survey data had a total population that differed significantly from Census 2011—suggesting shortcomings with the data collection at a basic level. These processes of re-enumeration occurred before the publication of the draft list when it was clear that the first attempt to collect data had serious problems. I later learned of a "Verification and Correction Module" that inserted a new step in the process.[9] The NIC would run the draft data file for each enumeration block through a new software to identify missing or inconsistent data in four areas for rural households (that is, caste, primitive tribal groups, legally released bonded labor, and manual scavengers) and in two areas for urban households (that is, main source of income and caste)[10] The data flagged by the NIC required additional scrutiny and the verifying officer at each charge center reviewed the data for these households, conducted the field verification (that is, reinterviewed households), made the required changes to the data, and then uploaded the corrected data to the NIC website. The NIC would then generate the draft list for public verification from the corrected files.[11] I am unclear whether the local officials in Bengaluru identified the blocks that required re-enumeration because of discrepancies in the total population between the initial SEC survey enumeration and the census, as shared by local enumerators, because of problems identified by the new verification software, or both. Given that the new data collection teams enumerated all of the households in these blocks, I suspect that the re-enumeration was related to a more systemic problem identified prior to the Verification and Correction Module. Either way, the process of re-enumerating blocks inserted an additional step that further lengthened the timeline for data collection.

Additional factors also contributed to an extended timeline for data collection. A Standing Committee on Rural Development responsible for overseeing the work of the MoRD identified problems in the implementation of the SEC survey and listed some causes for delays in the project in its 2016 Lok Sabha report:

> The Committee strongly criticize the mishandling of the overall planning and implementation of the [SEC survey] by the Government by way of shortening time limit from one year to six months for undertaking [the survey], starting the process only in June.... Various other factors that impeded the implementation of [the survey] including failure to conduct both Enumeration and Supervision concurrently within Tehsils/Districts, failure to provide internet connectivity for data transfer strikes, bandhs, frequent change in software as per State specifications etc. have also been outlined before the Committee. The Committee feel that all these issues put a big question mark on the overall project planning and implementation of [the survey] by all stakeholders including the nodal Ministry, the MoRD, and does not substantiate the claim of the MoRD about proper coordination between MoRD and States/UTs in the conduct of [the survey].[12]

The committee criticized the MoRD for its overall lack of preparedness and failure to effectively coordinate with state governments, who oversaw data collection operations. In addition, I will add that the process of recruiting and appointing DEOs often took longer than expected, and the tablets also had technical difficulties, along with the identified delays related to the data entry software and internet connectivity. An internal government report cited that 10–15 percent of the tablets had problems, extending the time it took to complete data collection and delaying the start of data collection in states that were scheduled to begin later, such as Bihar and Uttar Pradesh. These factors, along with data collection teams often taking longer than the allocated forty days to complete enumeration in their assigned blocks in urban areas, and the insertion of a whole new step for electronically reviewing collected data prior to public verification, delayed the completion of data collection. In total, the enumeration process that was slated to take about two months in most states took at least two years in most major states (see Table 6.1, columns b–c). On March 31, 2015, the MoRD reported that enumeration "was completed in 2,464,909 Enumeration Blocks (EBs) consisting of 99.46 percent of total EBs of all States/UTs."[13] Three years after the SEC survey data collection should have been completed across all regions of the country, data collection was almost complete.

The public verification process also took time. In several states, ranging from Andhra Pradesh to Punjab to West Bengal, the process took more than a year (see Table 6.1, column d). The MoRD reported in March of 2016 that 5.5 percent of enumerated households raised objections (13.6 million out of 246.4 million households across the country).[14] Based on the MoRD's reply to the Lok Sabha's Standing Committee on Rural Development, most of the claims and objections occurred in Bihar, Rajasthan, and Uttar Pradesh.[15] Upon the completion of the public verification process, the MoRD reported that 7.8 percent of rural households (14.1 million out of 179.97 million rural households) raised objections to draft lists when they were "put for public scrutiny before Panchayat and were displayed in Gram Sabha."[16] In a village, the posting of household information in the local office might lead to many if not most households verifying the data—especially since households understood it was connected to accessing BPL services. The public verification process was not well understood in urban areas. Both during the household interview stage, as discussed in Chapter 5, and during the public verification stage, most urban households did not even know that the SEC survey was taking place. The reliance of this stage for correcting errors seemed less effective in urban areas and stood in contrast to the MoRD's confidence that "since valid information on 100% households is now available, a distinct credibility is added to the final data which is now open for examination and deriving actionable meanings."[17] By early 2016, the government had completed its public verification process for SEC survey data.[18]

Processing the Caste Data

The government chose not to include the caste and religion data in the public verification process. The secretary of the MoRD explained in a memo to each state's chief secretary two months prior to the start of the survey that the

> caste and religion data of the person will also be collected during this census, though this information, unlike other information of this census, will not be made public and will be used by the Registrar General of India for statistical purposes only i.e., socio economic profiling of various castes in India.[19]

Table 6.1 Enumeration and Public Verification Timeline for SEC Survey in Major Indian States (2011–2015)

Major Indian State	(a) Date SEC Survey Enumeration Began	(b) Enumeration Period (days)	(c) Enumeration Period (years)	(d) Public Verification Period (days)
Andhra Pradesh	Nov. 22, 2012	1379	3 years, 9 months, 10 days	414
Assam	Feb. 08, 2012	1032	2 years, 9 months, 28 days	269
Bihar	Jan. 10, 2012	878	2 years, 4 months, 27 days	172
Chhattisgarh	Oct. 06, 2011	1156	3 years, 1 month, 29 days	298
Goa	Dec. 01, 2011	881	2 years, 4 months, 29 days	56
Gujarat	Jan. 28, 2012	836	2 years, 3 months, 15 days	569
Haryana	Nov. 27, 2011	1009	2 years, 9 months, 5 days	498
Himachal Pradesh	Oct. 22, 2011	1493	4 years, 1 month, 1 day	205
Jammu & Kashmir	Dec. 09, 2011	854	2 years, 4 months, 2 days	264
Jharkhand	Feb. 12, 2012	1149	3 years, 1 month, 25 days	469
Karnataka	Dec. 10, 2011	765	2 years, 1 month, 3 days	341
Kerala	Apr. 17, 2012	655	1 year, 9 months, 15 days	413
Madhya Pradesh	Dec. 27, 2011	1008	2 years, 9 months, 3 days	224
Maharashtra	Dec. 07, 2011	1101	3 years, 5 days	293
Odisha	Dec. 15, 2011	1394	3 years, 9 months, 24 days	259
Punjab	Nov. 17, 2011	1103	3 years, 7 days	562
Rajasthan	Nov. 28, 2011	1429	3 years, 10 months, 29 days	298
Tamil Nadu	Jun. 06, 2012	920	2 years, 6 months, 7 days	52
Uttar Pradesh	Jun. 18, 2012	1143	3 years, 1 month, 18 days	506
Uttarakhand	Dec. 15, 2011	778	2 years, 1 month, 16 days	165
West Bengal	Feb. 09, 2012	833	2 years, 3 months, 13 days	487

Source: Standing Committee on Rural Development, *Twenty Seventh Report* (Delhi: Lok Sabha Secretariat, August 31, 2016), 20.

Households did not have an opportunity to review the collected caste-wise data, unlike for other survey questions where that option at least theoretically existed, which perpetuated the problems observed in the collection of caste-wise data, as described in Chapter 5.

Every public statement about the project explicitly stated that caste-wise data would not be put into the public domain for verification, and the mandatory declarations at the end of the household interview confirmed the same. Each interviewed household agreed to be "giving its consent to place the information in public domain, except data on religion and caste/tribe."[20] Yet, in practice, the draft list of SEC survey data published if a household was SC or ST, as did the final list and subsequent BPL lists, since SC and ST status contributed to BPL identification. The project consistently promoted an ideology of castelessness where SC or ST identities at the household level were public, open to scrutiny, and included in verification processes while the caste identities of others were private. This dichotomy aligned with the broader ideology of castelessness that ran through the entire project. While the MoRD started releasing finalized data in July 2015—as described earlier—and the MoHUPA published provisional data in 2015 and final data in 2016, the caste-wise data remained on a government server, untouched.

Untouchable Data

No one wanted to take responsibility for the collected caste-wise data. It became increasingly unclear who was responsible for analyzing the data and the timeline for making the data public.

When asked about the caste data after the publication of provisional BPL data, the rural development minister Chaudhary Birender Singh said, "[T]his is the jurisdiction of DG (director general), Census. It is for him to decide what he thinks about it. Only he can satisfy your queries."[21] Senior MoRD officials repeatedly stated that the ORGI was responsible for analyzing and publishing the caste data.[22] As journalists reported at the time, "[T]he Rural Development Ministry has washed its hands of the matter while the [O]RGI too seems reluctant to claim ownership."[23]

In June 2015, when I visited the ORGI office in Delhi, a spokesperson shared that the caste data were not under the purview of the ORGI, and that the caste and tribe data would be given to the Ministry of Social Justice and Empowerment and the Ministry of Tribal Affairs as soon as the ORGI received the final data—which they expected to obtain by August 2015. He asserted that the Ministry of Social Justice and the Ministry of Tribal

Affairs were solely responsible for analyzing the data. When I pushed further, questioning why all the media coverage and official statements associated the caste-wise data with the ORGI, the spokesperson remained firm, explaining, "We received the data, but the analysis of caste-related data and matters have to do with the Social Justice Ministry, not us."[24] A reply by the Ministry of Home Affairs to questions posed in the Lok Sabha in early 2014 further confirmed this position. Three MPs asked whether the government had conducted a "caste based census in the country" and if so asked for the details of "the total population of persons belonging to SCs/STs and OBCs vis-à-vis the total population of each state in the country."[25] The minister of home affairs, R. P. N. Singh, replied that such an exercise was "being carried out" and that fieldwork was currently "being conducted by the respective State/UT governments." He went on to explain:

> Once the field work gets completed, the data of the [SEC survey] will be uploaded at the National Informatics Centre (NIC) Server. The Office of Registrar General and Census Commissioner, India would process the caste data and hand over the details of the castes/tribes returned in the [SEC survey] to the Ministry of Social Justice and Empowerment and the Ministry of Tribal Affairs for categorization and classification of the returns by the Expert Group to be constituted by the Central Government at an appropriate time in consultation with the Ministry of Social Justice and Empowerment and the Ministry of Tribal Affairs.

It appeared that the ORGI was to take the first pass at processing the data and then would share the compiled data with the Ministries of Social Justice and Tribal Affairs for analysis and tabulation. Census commissioner C. Chandramouli reiterated this point in July 2015 when he said, "We collected the data, now the nodal ministries in charge will process and release it. It would be done by the Ministry of Social Justice and Empowerment, and the Ministry of Tribal Affairs."[26] Similarly, journalists reported a senior Home Affairs official sharing:

> The caste census data was done by the enumerators sent by [O]RGI. The data is ready and it is a password-protected file in custody of NIC. [O]RGI has nothing to do with its processing and release, it is to be done by the Social Justice and Tribal Affairs ministries. [O]RGI only gave technical support.

In contrast to the position of the ORGI and Home Ministry, the Ministry of Social Justice and Empowerment remained notably silent. Social Justice secretary Anita Agnihotri refused to comment, while a senior official shared with reporters that the processing of caste-wise data had nothing to do with the ministry and clarified that it was the domain of ORGI, which reports to the Home Ministry.[27] While all units agreed that the caste-wise data had been collected as part of the SEC survey, no agency wished to be responsible for analyzing and publishing the data.

Politicians, the media, and members of the public continued to inquire about the fate of the caste-wise data in early July of 2015, after the MoRD published the provisional BPL data. Journalists Ruhi Tewari and Vijaita Singh put it this way in an article published on July 8, 2015:

> ... Various government authorities [have] been passing the buck to one another. Officials have been saying the caste data, compiled from the first such census since 1931, is part of a password-protected document kept in the custody of the National Informatics Centre, but no ministry is willing to accept its ownership.[28]

Politicians and media pundits speculated that the BJP government was intentionally not publishing the caste-wise data in advance of upcoming elections in Bihar. Others advocated for the government to release data for the effective implementation of affirmative action.[29] As more and more questions surfaced, the BJP government decided to act or potentially face political blow back in the upcoming elections.

Following a cabinet meeting on July 16, 2015, finance minister Arun Jaitley announced the creation of an expert committee headed by the vice chairman of the NITI Aayog, Arvind Panagariya, to analyze the caste data. Panagariya was a well-respected development economist from Columbia University who had joined Prime Minister Modi's government earlier in the year.[30] In his senior leadership role, Panagariya oversaw the new organization that replaced the Planning Commission. The Ministry of Home Affairs explained that the collected caste-wise data included 4.67 million caste names, consisting of "caste/sub-caste names, synonyms, surnames, clan/gothra names, phonetic variations, sections, sub-groups " that would have "to be classified by experts having domain knowledge on these issues."[31] The same statement disclosed that there had been 81 million errors in the data, and state governments had been notified about the errors. As of mid-July, state governments had corrected over 67 million errors, with now approximately 14 million errors remaining.[32] While the statement did not identify who detected the errors, all existing evidence pointed to the ORGI as the agency that initially processed the caste-wise data, and my interview with the ORGI spokesperson the month prior confirmed the same. At the same time, the ORGI continued to distance itself from the data, repeatedly stating that it had handed over the caste-wise data to the Ministry of Social Justice and the Ministry of Tribal Affairs.

Expert Committee to Analyze Data

The appointment of Arvind Panagariya to head the expert committee to analyze the caste-wise data meant that a government-appointed unit took public responsibility for publishing the data. Panagariya was not an expert on caste, but as a leading social scientist and a senior government appointee he could ostensibly lead a committee with experts on caste who could scrutinize and tabulate the data.

For months, there was little public information about the caste data. Approximately one year after Panagariya's appointment, a report from the Lok Sabha Standing Committee on Rural Development included a statement from the MoRD that discussed the completion of the SEC survey and also mentioned the caste and religion data:

> Canvassing of questions on caste and religion too were completed simultaneously. Registrar General and Census Commissioner of India, Ministry of Home Affairs has informed that the data has been examined and 98.87% data on individuals' caste and religion is error free. ORGI has noted incidence of errors in respect of 13,477,030 individuals out of total [SEC survey] population of 1,186,403,770. States have been advised to take corrective measures.[33]

The information from the ORGI meant that roughly 1.1 percent of the caste and religion data had errors. From the initial error count of 81 million shared the year prior, only 13.48 million errors remained—67 million or nearly 83 percent of the errors in the caste-wise data had been corrected.[34] Comparing the error rate of the caste-wise data in mid-2016 to other SEC survey data at the time of public verification (when 5.5 percent of enumerated households filed claims), the caste-wise data appeared less error-ridden and on their way to being made public.

Although information did not surface about the other members of the expert committee, answers to questions raised in parliament made it appear as if the committee was hard at work. In August 2016, an exchange in the Rajya Sabha signaled that the expert committee's analysis was ongoing. MP Veerendra Kumar asked the minister of home affairs three interrelated questions about whether the caste-wise data had been released, the reasons for delay, and the likely timeline for public release of the data.[35] The minister of home affairs, Hansraj Gangaram Ahir, replied:

> Sir, the caste related data, as enumerated in Socio Economic and Caste Census (SECC) has not yet been released. The Government has decided to constitute an Expert Group under the Chairmanship of Shri Arvind Panagariya, Vice-chairman, NITI AAYOG to classify and categorize the caste names returned in SECC. The Expert Group will be serviced by the Department of Social Justice and Empowerment.[36]

While the home minister did not specify when the expert committee was expected to finish their analysis and publish the caste-wise data, his response suggested that the expert committee was at work and supported by the Department of Social Justice and Empowerment.

Burying the Data

Time continued to pass without updates from the chair of the expert committee or any other members of the committee. Occasional anonymous reports to the media continued to reveal problems with the data. In the lead-up to assembly elections in Uttar Pradesh in early 2017, *The Asian Age* reported that the government did not want to release the data because of the number of OBC subcastes:

> According to highly placed sources, the [SEC survey] is learnt to have revealed that there are around 1,200 sub-castes under the OBC category alone, which has put the Centre under a quandary with the general feeling being that it would require a lot of time to streamline the "complex data." Also considering the sensitivity involved, any official publication could open a Pandora's Box for it. Sources privy to the developments told this newspaper that with 1,200 sub-castes falling under the crucial OBC category alone, the situation can easily be imagined if the data is released in its current format. "Every sub-caste will seek reservation under the OBC category, which is likely to create a mess. Many of these sub-castes have similar sounding surnames and gotras with minor variations. In some cases, there is only a difference of an alphabet and the sub-caste changes," said a government functionary on condition of anonymity. In such a situation, they added that there is a strong likelihood that details of the [SEC survey] concerning the actual figures related to caste may well be relegated to the back burner for the long haul.[37]

BJP political leaders foresaw several challenges if they published the caste-wise data, including that it might create considerable demand within the OBC quota and confusion whether groups identifying as part of the same OBC subcastes had been correctly grouped together. The article went on to explain that the expert committee chaired by Panagariya had over the last eighteen months "yet to make headway, thus giving an expression that the aim of the government was to only buy time, sources further informed."[38] The opinion that Panagariya's appointment was primarily to buy time seemed correct.

In mid-2017, Panagariya left his appointment in the Modi government and returned to his faculty position at Columbia University. The media began to conclude that it was the "end of the road" for the first caste "census."[39] Reporting by *The Hindustan Times* asserted that Panagariya's committee had submitted an initial report before he left.[40] Yet the committee never published a public document. Around the same time, anonymous reports also started to emerge from the government that the caste-wise data were unreliable. An October 2017 story from *The Times of India* stated:

> Also, there is scepticism in government echelons about the caste enumeration done in the [SEC survey]. A senior government source said no proof was sought for a respondent's claim of belonging to a particular caste and no attention was paid to technicalities like spelling of a caste which, with a minor difference, can change the social hierarchy. Many believe there is a link between the doubts about the [SEC survey] and the reason why the expert panel has not taken off.[41]

The expert committee's silence combined with anonymous reports about problems with the project's methodology fed the growing belief that the caste-wise data were unpublishable.

At the same time, state governments sought to access the caste-wise data. Maharashtra took the matter to court and argued that raw data from the SEC survey should be shared with state governments. In a reply affidavit to the Supreme Court on behalf of the union government in September 2021, the secretary of the Ministry of Social Justice and Empowerment, R. Subrahmanyam, wrote:

> The caste data has not been disclosed and has been kept with The Office of the Registrar General, India (ORGI) for various reasons, but primarily for the technical flaws that were noticed in the raw caste/tribe [SEC survey] data which makes it unusable as explained hereinafter. Therefore, the said data has not been made official for any purposes and cannot be mentioned as a source of information for population data in any official document.[42]

The secretary also confirmed that the ORGI had shared the data with the Ministry of Social Justice and Empowerment on a hard disk and that "due to several infirmities found in the aforesaid data as explained hereinafter" the cabinet had constituted "an Expert Committee under the Chairmanship of the then Vice Chairman [NITI Aayog] Prof. Arvind Panagariya."[43] The affidavit stated that "the other members in the committee were not named, and the Committee never met. As a result, no action has been taken on the data in the past 5 years."[44] The statement in a legal affidavit that the government never named the other members of the committee and that the committee never met was contradicted by an interview of Arvind Panagariya, which cited that the committee had released a preliminary report that had never been made public.[45] Either way, the secretary's point that a serious analysis of the caste-wise data had never taken place highlighted the government's long charade. The political leadership had preferred to simply bury the caste-wise data back in 2015 when the BPL data were published but, after coming under pressure to publish the caste-wise data by politicians in the lead-up to state assembly elections in 2016, had appointed Panagariya to head the expert committee and give the appearance that experts would publish the caste-wise data.

The secretary of the Ministry of Social Justice and Empowerment also outlined three different reasons why the caste data were flawed and "unusable" in the affidavit to the Supreme Court.[46] First, too many castes had been enumerated. In Maharashtra, the SEC survey enumerated 428,677 castes for a population of 105.3 million people. Nearly 84 percent of the population of Maharashtra was enumerated in one of the 2,440 castes that had a population of 1,000 or more.[47] The remaining 426,237 castes accounted only for 5.1 percent of the population, with each of these castes having a population of less than 1,000 people; the secretary wrote "that more than 99% of the castes enumerated had a population of less than 100 persons" in Maharashtra.[48] The SEC survey also enumerated 11.1 percent of the population of Maharashtra as "no caste."[49] The secretary argued that "every enumerator who visited each household spelt each caste separately" and cited the cases of the Mappilas in the Malabar region of Kerala wherein enumerators spelled the caste name "40 different ways," which resulted "into counting of 40 different/separate castes."[50] As such, the secretary's own statement suggested that by combining the numerous different spellings for each caste into one standardized spelling, there would be a significant reduction in the number of total castes. The secretary referenced that "the total number of castes during the first Census in India in 1931 was 4,147," while the most recent caste enumeration has found "46 lakhs [4.6 million] different castes. Assuming that some castes may bifurcate into subcastes, the total number cannot be exponentially high to this extent."[51] Given that the ORGI did not create caste lists for the entire population, an expected part of tabulating the caste-wise data should have been the commensuration of all responses that were neither SC, ST nor "no caste." In fact, the claim that 4.6 million castes were identified should not be a surprise for an open-ended question where no guidelines were provided to standardize caste names for consistency in spelling and other aspects of data commensuration for individuals falling within the state's "general" or "OBC" administrative categories.

Second, the process of reconciling the large number of enumerated castes with existing caste lists would be difficult. The secretary suggested that "there should have been given a drop-down menu for selection of the castes which could have made some consistent data available which can be relied upon."[52] This conclusion is consistent with the long-standing enumeration of SCs and STs in the census and historical processes of collecting caste data, when provinces created a list of castes found within the region prior to enumeration to streamline and standardize data collection. While creating regional caste lists requires an investment of time and resources prior to conducting the survey, it helps to ensure the commensuration of caste both within a state and across states and prevents a proliferation of caste names due to inconsistencies in spelling.

Finally, enumerators did not record caste data for many households. The secretary stated that data collectors enumerated more than 11 percent of the population of Maharashtra as "no caste."[53] The secretary made a similar point at the all-India level, stating that "in many cases, the concerned household has refused to divulge their caste and the enumerator has marked 'x' in the column of the caste showing either that the castes could not be determined or the household has refused to furnish the same."[54] Given that the MoRD website reported that only 1.43 percent of rural households were "no caste" in Maharashtra, did the "no caste" category largely consist of caste elites or religious minorities in urban areas, as described in Chapter 5?[55] Or did it reflect some other phenomenon? Understanding why data collectors in Maharashtra recorded 11 percent of the population as having "no caste" requires a geographic and socioeconomic breakdown of the caste-wise data and observations of on-the-ground data collection processes.

Across media interviews, legal affidavits, and public statements by politicians, the union government's position was clear: the caste-wise data were unreliable and could not be made public. The secretary concluded in the affidavit, "[T]here is not reliable or dependable caste-based census data available which can be the basis of any constitutional or statutory exercise like reservation in admission, promotion or local body elections."[56] The union minister of state for home affairs Nityanand Rai summarized the government's long-standing position in a reply to the Lok Sabha in 2022, "[A]s informed by the Ministry of Social Justice and Empowerment, there is no proposal to release the caste data at this stage."[57] The government had disappeared the caste data.

Looking Back and Ahead

Organizations and political leaders desiring caste-wise data have continued to demand access to SEC survey data, long after the government concluded that the caste-wise data were too unreliable to be publicly shared. Himanshu summarized this position in a 2023 op-ed published in *Mint*:[58]

> While the official explanation has been that the dataset is unreliable, this has been refuted by the registrar general office to the parliamentary committee. The data's error rate was found to be only 1.1%. In fact, this is not the only data on caste census that's neglected. There was a similar effort in Karnataka, which conducted its own caste census in 2017. However, even the results of that census have not been released.

Scholar-activists like Himanshu have pushed for a comprehensive analysis of the collected SEC survey data and its publication, despite its flaws, without success to date.

Following the collection of caste data, bureaucratic deflection continued to unfold during the processing and analysis of the data (see Table 6.2). First, as planned, the caste-wise data did not undergo a process of public verification (that is, other than SC and ST data). Since caste lists were not created prior to collecting the data (except for SC and ST data), there were a multitude of variations for many castes, and it became difficult to create meaningful "social equivalents" for 4.6 million entries. Second, the responsibility for analysis of the collected caste-wise data became seen as a liability. Once separated from the BPL data, no one wanted to take ownership of the caste-wise data. Confusion abounded over who was responsible for analyzing the data and the ORGI did not want to be in charge of finalizing and publishing the caste data or appointing an expert committee whose work it oversaw. Following political

Table 6.2 Institutional Strategies of Bureaucratic Deflection: Bury the Caste-wise Data

Strategy of Bureaucratic Deflection	Consequences of the Strategy
Caste-wise data did not undergo process of public verification	Missed opportunity for quality check and commensuration with public input
Lack of clarity regarding who was responsible for analyzing and publishing caste-wise data; agencies refused to take public responsibility for this stage of the process	Caste-wise data remained in a state of limbo, even after the rest of the SEC survey/BPL data were made public
Appointed the chair of an expert committee to analyze caste data—other committee members unnamed	Bought time and electoral advantage for BJP, as public believed publication of caste-wise data was imminent
Expert committee never published data; caste-wise data officially deemed unworthy of being published	Public has no access to collected caste-wise data; a detailed map of caste privilege and disadvantage unavailable

Source: Author.

pressure, the now BJP-led government created an expert committee and appointed a chair, but the committee (if it ever formed) never devoted serious resources and time to commensurate the data. Finally, the political leadership declared the caste data as unreliable and unworthy of public dissemination. This came from multiple unnamed government sources and then from the chair of the expert committee tasked with analyzing the data—after he left his senior government position to return to academia without the publication of a final report on the caste-wise data.[59]

The possibility of collecting caste-wise data now remains in future decennial censuses or with state governments, some of whom have taken up the mantle. The concluding chapter summarizes the findings of this case and examines more recent efforts to collect caste-wise data by state agencies as well as in the delayed 2021 Census.

Notes

1. Press Information Bureau, Government of India (Ministry of Finance), "Provisional Data of Socio Economic and Caste Census (SECC) 2011 for Rural India Released," news release, July 3, 2015, https://pib.gov.in/newsite/PrintRelease.aspx?relid=122963, accessed April 1, 2024.
2. Ibid.
3. The website for the project describes the overall coordination and specific roles in a similar way. Government of India, Ministry of Rural Development,

"SEC Census Homepage," https://secc.gov.in/homepage.htm, accessed July 14, 2023.

4. Interview with activist, February 22, 2013.
5. In addition, the data were made available on a government website during the public verification stage. Government of India, Ministry of Rural Development, "Socio Economic & Caste Census 2011 in Rural India," https://secc.gov.in/homePageUrban.htm accessed September 30, 2023.
6. Ibid.
7. Ministry of Rural Development, B. K. Sinha, "Socio Economic Caste Census Procedure, File D.R. No. Q-1615/04/2011AR(RD)," June 24, 2011.
8. B. K. Sinha, "Socio Economic Caste Census Guidelines (with Attachments)," June 24, 2011.
9. Government of India, Chief Economic Adviser, "Document No. H-1102/8/2010-CD(CEN)" (Ministry of Rural Development, October 26, 2012).
10. Ibid.
11. Ibid.
12. Standing Committee on Rural Development, *Twenty Seventh Report* (Delhi: Lok Sabha Secretariat, August 31, 2016), 9.
13. Fifteenth Lok Sabha Committee on Estimates Ministry of Rural Development, *National Social Assistance Program, 4th Report (2014–2015)* (Delhi: Lok Sabha Secretariat, April 29, 2015).
14. Standing Committee on Rural Development, *Twenty Seventh Report*,13.
15. Ibid., 13–14.
16. Government of India, Ministry of Rural Development, "SEC Census Homepage," https://secc.gov.in/homepage.htm, accessed July 14, 2023.
17. Ibid.
18. The National Capital Region of Delhi completed its verification of claims and objection on January 5, 2016. Standing Committee on Rural Development, *Twenty Seventh Report*, 20.
19. B. K. Sinha, "Socio Economic Census to Identify Below Poverty Line (BPL) Families," May 30, 2011.
20. Ministry of Rural Development, "SocioEconomic and Caste Census 2011 Questionnaire Rural."
21. Ruhi Tewari and Vijaita Singh, "Rural Development to Home to Social Justice, No One Wants to Own Undisclosed Caste Data," *Indian Express*, July 8, 2015.
22. Ibid.
23. Ibid.
24. Field notes, Delhi, June 29, 2015.
25. "Population of SCs/STs/OBCs, Lok Sabha Question No. 3925," Ministry of Home Affairs, February 18, 2014.

26. Tewari and Singh, "Rural Development to Home to Social Justice, No One Wants to Own Undisclosed Caste Data."
27. Ibid.
28. Ibid.
29. Ibid.
30. Panagariya was a strong supporter of market liberalization and Gujarat's recent growth trajectory.
31. "Government Defends Deferment of Caste Data; 8.19 Crore Errors Found," *The Economic Times*, July 28, 2015. "Caste Census Data Riddled with Errors," *Deccan Herald*, July 28, 2015.
32. Ibid.
33. Standing Committee on Rural Development, *Twenty Seventh Report*.
34. "Classifying Caste Census Data Could Be Panagariya's Toughest Job," *The Hindustan Times*, July 18, 2015.
35. "Rajya Sabha Question Number 2698," Ministry of Home Affairs, August 10, 2016.
36. Ibid.
37. "Longer Wait for Caste Census Information," *The Asian Age*, January 30, 2017.
38. Ibid.
39. Subodh Ghildiyal, "Is It End of Road for First Caste Census?" *The Times of India*, October 10, 2017; P. S. Krishnan, "Census Opportunity," *Frontline*, October 10, 2018.
40. Chauhan, "The Importance of Caste Census and Its Political Implications."
41. Ibid.
42. R. Subrahmanyam, *State of Maharashtra v. Union of India*, "Reply Affidavit on the Behalf of Respondent-UOI," No. Write Petition (Civil) no. 841 of 2021 (September 21, 2021), 3–4.
43. Ibid., 7–8.
44. Ibid., 8.
45. Chauhan, "The Importance of Caste Census and Its Political Implications."
46. Subrahmanyam, *State of Maharashtra v. Union of India*, "Reply Affidavit on the Behalf of Respondent-UOI," 3.
47. Ibid.
48. Ibid., 8.
49. Ibid.
50. Ibid., 11.
51. Ibid., 10. The affidavit cited over 4.6 million castes in total—consistent with other information released previously by the ORGI.
52. Ibid., 12.
53. Ibid., 9.
54. Ibid., 11.

55. Government of India, Ministry of Rural Development, "Caste Tribe Status of Households (Rural)," https://secc.gov.in/getCasteProfileRuralNationalReport.htm, accessed July 14, 2023.
56. Subrahmanyam, *State of Maharashtra v. Union of India*, "Reply Affidavit on the Behalf of Respondent-UOI," 12.
57. "No Plan to Release 2011 Caste Census Data: Govt," *The Economic Times*, July 26, 2022.
58. Himanshu, "We Require Reliable Data on Caste to Meet Our Aims of Social Justice," *Mint*, April 21, 2023.
59. In late 2023, the political leadership appointed Arvind Panagariya, the former NITI Aayog Vice Chairperson, and Jagdish N. Bhagwati, Professor of Indian Political Economy at Columbia University, chair of the 16th Finance Commission of India.

7

Conclusion

Commensuration in Brahmanical Institutions

In their 2008 article "A Sociology of Quantification," Wendy Espeland and Mitchell Stevens describe the immense bureaucratic effort required to quantify populations:

> Quantification requires considerable work, even when it seems straightforward.... Counting may seem like a simple act, but doing it on a large scale requires well-funded bureaucracies with highly trained administrators, especially if the counts are politically contested or "official"—and the two usually go together (Porter 1995). As many scholars have shown, producing a national census is an arduous undertaking (Anderson 1988; Derosières 1998; Loveman 2005).... Rigorous, defensible and enduring systems of quantification require expertise, discipline, coordination and many kinds of resources, including time, money, and political muscle. This is why quantification is often the work of large bureaucracies.... We often forget how much infrastructure lies behind the numbers that are the end product of counting regimes.

Chapters 3–6 trace the extensive infrastructure involved in the production of caste-wise data. In response to organizing by caste census advocates, the Congress-led government conceded to collect caste-wise data and then mobilized massive financial, human, and technological resources to do so. This infrastructure included, but was not limited to, executive bureaucrats who blocked the caste-wise enumeration in the census and pushed it into a BPL survey; MoRD administrators in Delhi who had been revamping the BPL survey for several years, who collaborated with the MoHUPA and the ORGI to coordinate and manage the combined BPL survey and caste-wise enumeration; state and local government officials that oversaw the implementation of the extensive ground operations for the survey; PSUs that managed electronic data entry operations; private firms that bid for, and secured, PSU-managed government contracts to staff and implement electronic data entry; charge centers—or the local hub for data collection operations, where data collectors uploaded interview data to a government server—and the people and technology that the staffed charge centers across the country; the government server hosted by the National Informatics Centre (NIC) that stored all collected household data; the entire training infrastructure for data collectors which prepared them to conduct household interviews; the hundreds of thousands of enumerators and DEOs who interviewed households, as well as the supervisors who monitored their work; the survey instrument, household listings, block-level maps, pre-loaded NPR data, tablets, software programs for electronic data entry and for data verification, and other material objects involved in enumerating households; the "public," or household respondents who provided responses to interviewers; the draft SEC survey data that were compiled and posted in local government offices throughout the country and on a government website; the local officials who oversaw the process of publicly verifying the draft survey data; the "public" that verified the data and reported errors; the MoRD, MoHUPA, and ORGI staff who reviewed the draft data and MoRD and MoHUPA staff who analyzed and tabulated the finalized data for BPL identification; the ORGI staff and state and local government employees who identified and corrected errors in the collected caste-wise data; the finalized BPL survey data publicly released by the MoRD and the MoHUPA between 2015 and 2016 (which did not include caste-wise data); advocacy by caste census advocates for publication of the collected caste-wise data; the BJP-led government's appointment of a chair to lead an expert committee to analyze the caste-wise data under the jurisdiction of the Ministry of Social Justice and Empowerment; the decision by the executive bureaucracy to deem the collected caste-wise data as unreliable and unfit for public dissemination and policymaking; and the buried and publicly inaccessible caste-wise data. The infrastructure for the survey project was vast and required considerable resources and coordination.

Understanding why the politically contested caste-wise enumeration "failed" requires attention to the embeddedness of Brahmanism (that is, the ideological structure of caste) throughout the administrative bureaucracy and survey-making infrastructure. Efforts to count caste in Census 2011 were unsuccessful due to extensive resistance from within the executive bureaucracy that blocked a project that would have made visible caste privilege and power. Once the caste-wise enumeration was inserted into another state project, the bureaucracy deflected its responsibility, and ideologies of castelessness,

Hindutva, and Brahmanical patriarchy structured the collection and commensuration of caste-wise data.

In this concluding chapter, I start by summarizing the mechanisms of bureaucratic deflection that unfolded within the executive bureaucracy to prevent a successful nationwide enumeration of caste. I then briefly discuss the actions of other actors that supported bureaucratic deflection. Focusing on a situation that continues to develop as I finish this book, I explore the case of the recently published caste-wise data from the north Indian state of Bihar to consider how bureaucratic deflection might be challenged. Finally, I explore how an ideology of castelessness precludes the possibility of dismantling Brahmanism and caste hierarchy in the early twenty-first century and situate the book's findings more broadly.

Bureaucratic Deflection Revisited

Attempts to block a caste-wise enumeration in Census 2011 by executive bureaucrats started as soon as the first public inquiries and efforts to push for a change in census policy began between 2008 and 2009. The early public responses of political leaders reflected the views of the ORGI—describing a potential caste-wise enumeration as too "cumbersome," likely to "jeopardize the very process of Census Operation" and affect the "basic integrity of the Census data," including "the fundamental population count."[1] Subsequent mobilization by caste census advocates created a broad base of support by May 2010, which helped to secure a public concession from the Congress-led political leadership to collect caste-wise data in Census 2011. In the aftermath of this concession, census bureaucrats redoubled their efforts to keep a caste-wise enumeration out of the census. In the months that followed, the political leadership maneuvered back and forth, promising to collect caste-wise data in the census but also identifying several other alternatives. Looking back, the authority of the ORGI and Home Ministry was at the foundation of the bureaucratic deflection that unfolded. Although it appeared that the political leadership had conceded to the demands of caste census advocates, the ORGI and Home Ministry in fact succeeded. The May 2010 concession resulted in neither the collection of caste-wise data in Census 2011 nor the publication of caste-wise data through another state project. B. R. Ambedkar's critique of the absence of caste-wise data from Census 1951—the first census of independent India—offers a preview to more recent processes of bureaucratic deflection:

> The last census omits altogether the caste tables which had been the feature of the Indian census ever since its birth. The Home Minister of the Government of India who is responsible for this omission was of the opinion that if a word does not exist in a dictionary it can be proved that the fact for which the word stands does not exist. One can only pity the petty intelligence of the author.[2]

The broader context of Ambedkar's comments related to his frustration in wanting to cite census data on caste to describe caste hierarchy and its effects in order to inform government policies (in this instance on the formation of linguistic states). Ambedkar argued that not enumerating and publishing caste-wise data as part of the census was a naïve approach to annihilating caste. A clear understanding of how caste hierarchy intersected with socioeconomic resources, occupation, gender, regionality, and other identities that shaped access to resources and dignity was important to develop and implement policies to challenge caste hierarchy. More than half a century later, the failure to include a caste-wise enumeration in Census 2011 reveals how bureaucratic deflection continues to reinforce an ideology of castelessness in state projects and protect the invisibility of Brahmanism.

The initial steps of bureaucratic deflection that unfolded between May 2010 and May 2011 repeated previous efforts to block a caste-wise enumeration in the census as described in Chapter 2. Executive bureaucrats tried to shift responsibility for the caste-wise enumeration from the central government to state governments (or state-level commissions); steered the debate away from a social justice framing (that is, documenting caste-based inequalities) to a discussion of methodological challenges; and compartmentalized the caste-wise enumeration to a peripheral and/or under-resourced unit in the bureaucracy. Cabinet members and other senior political leaders deployed these familiar maneuvers in the lead-up to Census 2011, even as caste census advocates fought back, as detailed in Chapter 3. While the executive leadership was the public face of these decisions, they had the stamp of senior bureaucrats from the ORGI and Home Ministry. In addition, several innovative steps of bureaucratic deflection followed, which I summarize in Table 7.1.

Newer mechanisms of bureaucratic deflection also emerged between May 2010 and May 2011, prior to when the "caste census" merged with the BPL survey. First, the political leadership secured operational latitude on how to collect the caste-wise data. The group of cabinet ministers tasked with implementing the "caste census" obtained written leeway from political parties which allowed them to push the caste-wise enumeration into another state project, as desired by the ORGI. The Congress-led executive bureaucracy put

Table 7.1 Bureaucratic Deflection in the Executive Bureaucracy: The Case of the 2011 Caste Count

To count or not to count caste
Decentralized responsibility from central to state governments (or state-level commissions)
Shifted debate from focus on documenting caste-based inequalities to decisions framed as purely technical considerations
Compartmentalized project to a peripheral and/or under-resourced unit in the bureaucracy
Ideology of castelessness promoted by the executive bureaucracy and other powerful state and non state actors
Pushed caste count out of the 2011 decennial census
Silenced voices of "non-experts" and homogenized different types of caste consciousness
Created operational latitude for project while securing support from opposition political parties
Put forward a series of alternative plans until executive bureaucrats successfully inserted the project in the BPL survey
Embedded caste count in an alternative state project
Decentralized responsibility for publicity; left responsibility to lower tiers of government so no publicity when survey begins
Data collection teams forced to play additional role as consent-builders, since households did not know about the survey
Implemented real-time electronic data entry, with insufficient time to scale up and plan for massive change in operations
Failed to create a list of castes prior to survey
Did not include "OBC" and "general" administrative categories in first caste-related question
Added "no-caste" answer option to first caste-related question which became an opt-out for caste-privileged groups
Lack of training for interviewers on how caste intersects with religion and how to enumerate caste of religious minorities
Buried collected caste-wide data
Caste-wise data did not undergo public verification, unlike other SEC survey data
Government agencies refused to take public responsibility for finalizing the caste data
Chair of expert committee appointed but unclear whether committee met; no final committee report published
Caste-wise data officially deemed unworthy of being published; public has no access to collected caste-wise data

Source: Author.

forward three different plans for the "caste census"—distinct from the 2011 census. Second, the voices of "non-experts" became silenced after executive bureaucrats shifted the debate from social justice concerns to largely technical considerations. Census bureaucrats dictated decisions related to methodology and disproportionately shaped the decision-making of top political leaders. Curbing the contributions of "non-experts" meant that powerful voices experienced in challenging castelessness and Brahmanism were absent in how and why to collect caste-wise data in Census 2011 and in subsequent bureaucratic processes and decisions. Instead, political leaders, bureaucrats, and the media began to homogenize different types of caste consciousness—inattentive to, or even masking, differences in political commitments to social justice and anti-caste organizing.[3] The political leadership eventually punted the caste-wise enumeration out of the ORGI's primary jurisdiction and into a revamped BPL survey, which came under the primary jurisdiction of the MoRD. The MoRD had a poor and inexperienced track record of administering nationwide surveys.

During the combining of the caste-wise enumeration and the BPL survey, five additional mechanisms of bureaucratic deflection transpired. First, the designers of the SEC survey chose to decentralize responsibility for publicity from the center to state and local governments. In doing so, they tasked units, already responsible for developing extensive ground operations, with the additional task of designing and rolling out a publicity campaign. This stood in contrast to the central government's extensive publicity campaign for the decennial census, which usually started more than a year prior to the main round of enumeration to inform the public and ensure widespread participation. State and local governments occupied with building the extensive SEC survey infrastructure for data collection had difficulty prioritizing publicity at the outset of the project. As such, most households—particularly in urban areas—had little idea that a BPL survey or caste-wise enumeration was underway. The role of the data collection teams became even more significant in places where publicity was lacking, as they had to create legitimacy for the project. In these locations, enumerators and DEOs became consent-builders to secure participation in the project. The need for data collection teams to play a lead role in consent-building affected how they administered the survey. Data collectors sought to make sure that respondents felt comfortable participating in the survey and avoided asking or drawing attention to questions that would create tension during the interview. For example, I observed some data collectors offering up the "no caste" option to a subset of elite, well-to-do households while other interviewers did not ask religious minorities about caste.

A second mechanism of bureaucratic deflection that arose during the planning of the SEC survey was the decision to implement real-time electronic data entry, with little time to scale up and plan for the project. The related plan to create a two-person data collection team—an enumerator to ask questions and a DEO to record household responses using a tablet—slowed down the completion of the survey considerably (that is, from the planned six months to more than two years in most places). Staffing delays in recruiting, hiring, and training DEOs and postponements in the availability of tablets and in the replacement of faulty tablets added lengthy delays to the data collection process. Unintendedly, these decisions, along with the structure of the data entry program, also elevated the role of the DEO to the position of lead interviewer in many instances. The lack of flexibility to move freely throughout the electronic questionnaire, as well as the importance of securing buy-in for the project—for which the tablet and preloaded NPR data became useful signs of official government work—led the DEO to often start and continue the interview, despite the fact that the purely technical training for DEOs neither introduced them to the content of the survey nor taught them how to conduct interviews.

Two additional mechanisms of bureaucratic deflection arose during the planning stage of the survey connected to the design of the caste-related questions. First, the historical knowledge acquired by enumerating caste in colonial censuses was ignored.[4] The senior administrators in charge of finalizing the caste-related questions—that is, ORGI bureaucrats—did not add the administrative categories of "OBC" and "general" to the first caste-related question with categorical answer options. In the second caste-related question, they limited their use of caste lists to SC and ST lists, and there was no effort to have enumerators (or supervisors) postcode the caste-wise data or standardize spellings of caste names. The long-standing practice of enumerating SCs and STs in the decennial census relies upon such pre-constructed lists, and builds upon the lessons learned across successive censuses in the colonial period. These decisions ensured that the commensuration of caste—that is, "the expression or measurement of characteristics normally represented by different units according to a common metric"—would be extremely difficult, as the two caste-related questions created open-ended responses for 77.5 percent of the population (that is, all "non-SC/ST" individuals). For a population of 1.2 billion individuals, the survey was therefore designed to capture open-ended caste data for 930 million individuals (minus those who responded "no caste," which I discuss later)—a decision that eventually generated over 4.6 million different responses. Arvind Panagariya—the chair of the expert committee appointed to analyze the caste-wise data in 2015—critiqued the caste questions in a 2021 interview with the *Hindustan Times*, stating that "there were no clear fields in the enumeration charts" and little consideration had been given to how the data could be used before the enumeration started.[5] The inattention that Panagariya describes—whether it was due to a lack of consideration or intentional—led to data with serious obstacles to commensuration. In contrast, the use of existing OBC lists and the creation of caste lists for the "general" category would have allowed for a streamlined process of interviewing households and aggregating and tabulating the collected caste-wise data; these steps would have required additional preparatory work and field testing of the caste lists.

Another decision made about the caste-related questions—and the fourth mechanism of bureaucratic deflection that emerged during the planning of the SEC survey—was the inclusion of a "no caste" answer option. The "no caste" answer option in practice reinforced an ideology of castelessness. From my observations, enumerators offered up the "no caste" answer to a subset of social elites who readily acquiesced when given the option, and in doing so obscured caste privilege. I also observed some enumerators recording Muslims as having "no caste" without asking the households themselves. This pattern relates to the lack of understanding among data collectors of how caste and religion intersect for religious minorities in India, as I will discuss next. In general, the inclusion of the "no caste" answer option appeased those who did not support a caste-wise enumeration, thereby reinforcing castelessness and Hindutva in the survey design and data—and provided an out for the successful enumeration of religious minorities. Whether my observations from Bengaluru align with broader nationwide trends for the "no caste" responses is difficult to verify, as the collected data have been buried.

A fifth mechanism of bureaucratic deflection that developed during the planning phase of the survey was the absence of a comprehensive training program on the caste-related questions. The lack of information in the training materials and in-person trainings about how caste and religion intersect (except for SCs and STs) meant that enumerators had no guidance on how to enumerate the caste of religious minorities. The failure to provide clear instructions on the caste-wise enumeration of 200 million Muslims as well as Christians and other religious minorities led to incommensurable data—with some enumerators recording sect or denomination instead of caste, others recording "no caste," and some documenting caste. The instructions for enumerators were also insufficient in how to approach the enumeration of privileged castes—who might be hesitant to discuss caste given their caste-based privilege. The enumerator instructions drew heavily upon the past training materials for the decennial census, which focused on the enumeration of SCs and STs and did not sufficiently adjust for the caste-wise enumeration of the entire population. In addition, DEOs received only minimal technical training to operate equipment and had no preparation for the caste questions or any other interview questions. This proved particularly problematic when the DEO took on the role of the lead interviewer or played an important role in interviewing households. Many households did not have prior knowledge about the questions or answers options—especially where publicity for the survey was absent—so their knowledge of survey questions and answer options came from what was shared by data collectors, who also had limited and inconsistent understandings of how caste and religion intersected across religious communities.

The final three steps of bureaucratic deflection unfolded after the collection of caste-wise data. First, the caste-wise data did not undergo public scrutiny. While the government posted the bulk of the SEC survey data for the public verification process, the review of caste data was limited to SC or ST status, as these data informed BPL identification. This approach prevented the public from seeing and making corrections to the recorded caste-wise data and meant that errors and inconsistencies in data, including differences in spelling of the same caste name and variations in the level of classification (for example, Brahman versus Iyengar), would remain.

Government agencies also refused to take public responsibility for analyzing the caste-wise data. This was a powerful maneuver that unexpectedly surfaced at a late stage. The political leadership had designated the ORGI as the nodal agency for the caste data, but the ORGI asserted that the Ministry of Social Justice and Empowerment was responsible for data analysis, as Chapter 6 details. No entity wished to oversee the analysis and circulation of the collected caste-wise data. Caste census advocates demanded the release of the caste-wise data following the MoRD's publication of the other SEC survey data in 2015. In response, the BJP-led political leadership appointed a chair for an expert committee to analyze the caste-wise data. While it remains unclear whether the committee ever met, the committee never released a public report.

Finally, senior government officials decreed the caste-wise data unworthy of dissemination and use. This opinion spread first through unofficial channels and later in response to media inquiries, parliamentary questions, and legal petitions pressuring the union government to publish the caste-wise data. Usually, when problems exist with "raw" survey data, experts clean up inconsistencies and correct errors as much as possible (a process that the ORGI oversaw) and then publish the results with a stipulation of the conclusions that can be drawn given the data quality and related concerns. The government's burial of the caste-wise data and decision to make it unavailable to the same public that funded the project and provided the data is unprecedented in a nationwide survey. However, the data being low quality and unusable is now accepted both in and beyond India.[6]

While the main story of *bureaucratic deflection* unfolded in state agencies, organizations outside of the state also played a crucial supporting role. One of the most powerful external responses came from the *meri jati Hindustani* campaign, which opposed the documentation of caste hierarchy. It helped reframe the debate by creating a nationalistic rallying cry that shifted the public conversation away from the durability and pervasiveness of caste-based inequality to supporting (an uncritical) nationalism. The campaign's nationalism homogenized different types of caste consciousness and promoted false equivalencies between caste consciousness that seeks to dismantle caste hierarchy and ideologies that undergird and reinforce caste hierarchy and power. Journalists, public intellectuals, and private citizens formed the campaign in Delhi, but celebrities and academics helped to amplify their voices. Only a limited group of academics and researchers offered critiques about why the data could be useful.[7]

Similarly, while the Supreme Court has repeatedly sought updated caste-wise data when making decisions related to reservations, it ruled that intervening in this case was outside its jurisdiction. The court derided the decisions of state high courts that sided with petitions to collect caste-wise data in the decennial census. In a 2014 decision, the Supreme Court called the move "a colossal transgression of power of judicial review" and argued against a lower court decision to require the collection of caste-wise data in Census 2011:[8]

> The [Census] Act has conferred power on the Central Government to issue Notification regarding the manner in which the census has to be carried out and the Central Government has issued Notifications, and the competent authority has issued directions. It is not within the domain of the Court to legislate.... [T]he courts are not to plunge into policymaking by adding something to the

> policy by way of issuing a writ of mandamus. There judicial restraint is called for remembering....[9]

The Supreme Court emphasized the executive's authority in deciding census policy. Thus, the courts became an ineffective avenue to advocate for change in decennial census policy. Still, a more recent decision by the Bihar High Court, which the Supreme Court did not challenge, has supported the collection of caste-wise data in a state survey, and advocates for the ORGI to do the same—as previous state high court decisions in south India have recommended.[10]

This book traces how bureaucratic deflection prevented one effort to make caste-based power visible. At every stage of the failed "caste census," from if and how to collect the data to whether to publicly disseminate the data, the executive bureaucracy ensured the failure of the project—across both Congress and BJP administrations. The observed phenomenon likely applies to a broader set of cases that seek to name caste privilege and dismantle gendered caste hierarchy, as well as other systems of oppression that have structured the formation of bureaucratic expertise and the composition and culture of the executive bureaucracy.

Reject Castelessness, Name Caste Power

The central government has refused to change census policy on caste, and its position remains the same as it makes plans to conduct the delayed census of 2021. Public comments in 2018 by officials in the BJP-led government indicated that the ORGI would collect data on OBCs in Census 2021. Yet, by the following year the administration realigned its position with the post-independence Congress leadership.[11] Once re-elected to a second term in office in 2019, the Modi government did not see any reason to follow through on its earlier promise made by the home minister and census commissioner. In July 2019, just before the pre-test of Census 2021 began in select regions, census officials publicly shared that the pre-test did not collect data on OBCs and that it was unlikely that Census 2021 would do so. The ORGI published the questions to be included in the household listing phase of Census 2021 in the *Gazette of India* on January 9, 2020 (as per the Census Act of 1948 and Census Rules of 1990). The caste-related questions were limited to the identification of "whether the head of the household belongs to Scheduled Caste/Scheduled Tribe/Other."[12] Soon after, the COVID-19 pandemic delayed the start of the household listing phase of the census, which should have taken place between April and September 2020, and the main round of the population enumeration, which was expected to occur in early 2021. Replying to a written question in the Lok Sabha, the minister of state for home affairs Nityanand Rai confirmed in July 2021 that "the government of India has decided as a matter of policy not to enumerate caste-wise population other than SCs and STs in census."[13] In 2022, the government amended the Census Rules of 1990 to allow for electronic data entry and self-enumeration—two significant changes that will transform data collection operations for the decennial census.[14] The Census of 2021 is now scheduled to begin in late 2024, following parliamentary elections, but without a caste-wise enumeration. As such, census data will remain unavailable to inform reservation policies for OBCs and the more recent implementation of a 10 percent reservation in government jobs for "economically weaker sections" (EWS) of the "general" category (or caste-privileged groups), as well as other government programs.[15] Caste-wise data from the census could be a powerful tool to challenge the exclusion of caste-oppressed groups from the EWS quota.

State governments continue to fill the void, as they have done in earlier periods, with differential outcomes. As of January 2024, the case of Karnataka remains a cautionary tale: caste-wise data collected by state governments can also be buried. Karnataka, which is home to one of the oldest systems of affirmative action in the country, implemented a caste-wise survey in 2014 following years of on-the-ground organizing. Karnataka's BCC, a state organization that has produced seminal reports documenting caste hierarchy and implementing reservations, oversaw the caste-wise data collection. In the end, the political leadership of the state buried the caste-wise data such that the collected data have not been published. The state's political leadership has generally come from groups that have disproportionately benefited from the long history of OBC reservations for non-Brahmans; leaders were ostensibly worried the publication of caste-wise data might be unpopular among comparatively well-to-do OBC communities and risky for their political careers.

Bihar has experienced a different outcome. The state government began the collection of caste-wise data in early 2023. The survey consisted of seventeen questions that collected socioeconomic information on income, employment, housing, level of education, work-related migration, vehicle ownership, and laptop ownership. The state created a list of 215 castes that enumerators used to record caste-wise data during household interviews. Unsurprisingly, several groups tried to stop the caste-wise data collection through the courts.[16] After numerous petitions to stop the survey proved unsuccessful, the state government published a summary of the caste-wise data in early October 2023, and the published data have already had profound reverberations. The summary caste table identified two categories of OBCs—distinguishing between the comparatively privileged communities and those who have been systematically discriminated against and unable to access the benefits of reservation, including the identification of caste-oppressed Muslims. The data offer a clear challenge to castelessness, which systematically masks the pervasiveness and magnitude of caste-based inequalities and the concentration of power and resources among privileged castes.

Following the publication of the preliminary data in October 2023, the state published a detailed report a month later on the caste-wise population of Bihar. The collected data have influenced revisions to affirmative action quotas in government jobs, higher education, and elected positions. The state assembly of Bihar passed two bills to expand caste-related reservations from 50 percent to 65 percent in November 2023, and other efforts to use the caste-wise data to inform policy are underway.[17] One of the ongoing challenges of using caste-wise data to shape policymaking will be to ensure the priorities of the most marginalized communities are not secondary to electoral calculations that often favor comparatively well-to-do and politically organized communities.

Similar to the unfolding situation in Bihar, a caste-wise enumeration in the decennial census could provide important insights to make certain that those systematically excluded from resources and opportunities benefit from redistributive and representational policies and their expansion. However, this story of bureaucratic deflection also highlights the need for supposed "non-experts" (from the perspective of census bureaucrats) to oversee and monitor the design and execution of a caste-wise enumeration in the census. From the design of the survey instrument to the creation of caste lists, a committee of anti-caste scholar-activists, including those in support of a caste-wise enumeration and those who have opposed it but have a political commitment to annihilate caste hierarchy, should inform the entire census-making process. Given the power of castelessness and Hindutva, external oversight will be key to the collection, commensuration, and publication of meaningful census data on caste that will make caste hierarchy and privilege visible.

At the same time, while caste-wise data from the census are important, they are just the start. As Satish Deshpande argues, caste-wise data will provide a collective self-portrait of the country to inform policies and programs to dismantle caste hierarchy.[18] Then, the hard work still remains of using the data for the design and expansion of social justice policies and programs. In a recent publication, "Counting Caste: A Step towards Radical Anticasteism," political scientist Vivek Kumar Singh states that "the caste census is a necessary but not a sufficient condition for practicing anti-casteism. Conducting the caste census will shift the gaze away from Dalits, and lead to the recognition of the problem of caste, a problem of the whole society."[19] Kumar argues that the caste census is an important step in providing the "anticaste Ambedkarite discourse with the much-needed data to bring out the effect of caste in the material sphere and include the aspects of both representation and redistribution."[20]

The failed caste census follows a broader trajectory of the dominant political parties supporting the institutionalization of castelessness in independent India. The executive bureaucracy generally operates by obscuring caste privilege and approaching caste as a problem of oppressed groups. The relationality of caste remains largely unacknowledged, and the systems that perpetuate caste are depicted as archaic remnants of the past. Instead, we know from scholarship in critical caste studies that caste hierarchy is continuously reproduced in the most modern of spaces, from high-tech companies in Silicon Valley to elite technical educational institutions to overseas scholarship and professional networks that help create future judges, CEOs, journalists, and faculty.[21] Castelessness perpetuates the myth that caste hierarchy is residual and that globalization and liberalization are dismantling it to create a system where anyone willing to work hard can find success. Castelessness also minimizes the institutionalized nature of Brahmanism. The pervasiveness of Brahmanism neither hinges on individual intentions to discriminate nor requires that individuals intentionally act to protect their own caste privilege. Individuals who are in support of organizational norms of meritocracy and efficiency—without attention to structural advantages based on caste, religion, gender, disability, and class (among other factors)—may help to strengthen Brahmanical ideologies and outcomes. The conscious and unconscious dimensions of Brahmanism are supported by multiple ideologies, normative frameworks, and discursive registers. The institutionalization of Brahmanism means that policies, practices, and norms not explicitly about caste may help to reproduce caste power. Many of the new mechanisms of Brahmanism appear unrelated to caste—and make no explicit reference to caste—but instead promote seemingly caste-neutral ideals of nationalism and expertise. These mechanisms are particularly resilient ways to appear consistent with democratic norms while reproducing oppressor-caste power.

The embeddedness of Brahmanical ways of thinking within powerful institutions creates barriers to the development and implementation of anti-caste policies and liberatory spaces, even as groups organize and seek to challenge castelessness, make Brahmanical power visible, and resist it. A caste census could provide crucial data for the challenging work of building coalitions, dismantling caste privilege and power, and centering the voices and demands of groups systematically devalued and ignored by gendered caste oppression—all with the goal of annihilating caste.

Notes

1. Maken, "Document No. H-1102/8/2010-CD(CEN)."
2. Ambedkar, "Thoughts on Linguistic States," 167.
3. Roberts, "Sources of Commitment to Social Justice."
4. During colonial censuses, census officials learned from regional administrators that creating regional lists of castes prior to data collection was a social process that helped to commensurate complex localized systems of caste hierarchy within and across regions.
5. Chetan Chauhan, "The Importance of Caste Census and Its Political Implications," *Hindustan Times*, February 26, 2021.
6. For example, the Pew Research Center, a nonpartisan Washington DC-based think tank, stated in an entry entitled "Measuring Caste in India" on its blog *Decoded* that a survey took place in 2011 "with the aim of obtaining more accurate numbers on lower-castes in India, including OBCs, but this data has not been released due to data quality concerns." Kelsey Jo Starr and Neha Sahgal, "Measuring Caste Data in India," *Pew Research Center Decoded* (blog), June 21, 2021.
7. As discussed in Chapter 3, there was a small and vocal group of academics that offered a counternarrative. See Japhet and Gowda, *Caste Census towards an Inclusive India*.
8. *Dipak Misra, Census Commissioner and Others v. R. Krishnamurthy*, No. Civil Appeal No. 9996. (Supreme Court of India November 7, 2014).
9. Subrahmanyam, *State of Maharashtra v. Union of India*, "Reply Affidavit on the Behalf of Respondent-UOI," 24–25.
10. Abraham Thomas, "Pleas against Caste Census in Bihar Merely for Publicity: Supreme Court," *The Hindustan Times*, January 21, 2023.
11. Complementing the work of the Rohini Commission, which was tasked with examining the sub-categorization of OBCs, the Home Ministry announced in 2018 that OBCs would be enumerated in Census 2021. In a meeting chaired by the home minister on August 31, 2018, the minister of state for home affairs Kiren Rijiju, home secretary Rajiv Gauba, census

commissioner Saliesh, and other senior officers of the ORGI and Home Ministry reviewed plans for Census 2021 and decided to collect data on OBCs and improve the design of data entry and processing. A spokesperson from the Home Ministry announced, "Census 2021 will be finalised in three years after conducting the census instead of seven to eight years now. It is also envisaged to collect data on OBCs for the first time." This announcement was significant because, for the first time, prior to the implementation stage of the census, the Home Ministry and ORGI took the lead and agreed to collect caste-wise data in the census and with sufficient advanced notice to effectively execute the policy change. Rahul Tripathi, "OBC Count to Be Part of Census 2021, 3 Decades after Mandal Commission," *The Indian Express*, September 1, 2018; "In a First, Government to Collect Data on OBCs in Census 2021," *The Financial Express*, September 1, 2018.

12. Ministry of Home Affairs, Office of the Registrar General & Census Commissioner, India, "Notification S.O.120(E) [Dated: 1/7/2020], [F. No.9/7/2019-CD(CEN)3]," *The Gazette of India*, January 9, 2020.
13. "No Caste Enumeration Other than SC and ST in Upcoming Census, Says Govt," *The Wire*, July 20, 2021.
14. Ministry of Home Affairs, Office of the Registrar General & Census Commissioner, India, "REGD. Notification D. L.-33004/99," *The Gazette of India*, March 11, 2022.
15. This policy initially put forth by the Congress Party decades earlier was implemented by the BJP-led government in 2019 by the 103rd Constitutional Amendment and upheld by the Supreme Court in 2022.
16. A caste-based survey started in Bihar with the household listing phase in January 2023. On January 20, 2023, the Supreme Court formally refused to consider petitions challenging the Bihar government's notification on June 6, 2022, to collect caste-wise data and suggested there was no merit in the pleas and called them "publicity interest litigation." With much of the survey complete, the Patna High Court placed an interim stay on the caste-based census on May 4, 2023, while it sought time to hear a petition that challenged the caste-based census. The high court said in its initial review that "[p]rima facie, we are of the opinion that the State has no power to carry out a caste-based survey, in the manner in which it is fashioned now, which would amount to a census, thus impinging upon the legislative power of the Union Parliament." It directed the government to immediately stop its data collection and secure the collected data. The Government of Bihar then filed a petition with the Supreme Court challenging the interim order of the Patna High Court that stayed the Bihar Government's decision to conduct the caste census, but the Supreme Court refused to interfere, citing that the Patna High Court would review the issue in early July. In early July, the Patna High Court heard five days of detailed arguments in

which the lawyer appearing on behalf of the petitioners opposed to the caste census argued that the state survey is in violation of the Indian Constitution according to which only the central government is authorized to conduct a census. Chief Justice K. Vinod Chandran and Justice Partha Sarthy on August 1 upheld the legitimacy of the survey, and in early October 2021 Bihar published the findings from its first caste census. This case highlights the predicament of advocates for the caste census who have been blocked by the central government executive bureaucracy for over seventy years, but who also face challenges while collecting and publishing the data via state-level surveys. Thomas, "Pleas against Caste Census in Bihar Merely for Publicity: Supreme Court"; Sohini Chowdhury, "Supreme Court Refuses to Interfere with High Court's Stay on Bihar Caste Survey as HC Is Set to Hear Main Matter in July," *Live Law*, May 18, 2023; Sparsh Upadhyay, "Patna High Court Reserves Verdict in Pleas Challenging Bihar Govt's Decision to Conduct Caste-Based Survey in State," *Live Law*, July 7, 2023; "Patna High Court Dismisses Petitions against Bihar Caste Census," *The Wire*, August 1, 2023; Kunal Purohit, "How a Landmark Caste Census in India Threatens Modi's Grip on Power," *Al Jazeera*, October 4, 2023; Anirban Guhaand Roy and Subhash Pathak, "Decoding Findings of Bihar's Caste Survey Report," *The Hindustan Times*, November 8, 2023.

17. "Bihar Passes Bills Raising Caste-Based Quota from 50% to 65%, Taking Total Reservation to 75%," *The Wire*, November 9, 2023; Shivani Vij and Shrutanjaya Bhardwaj, "Can Bihar Increase Its Reservation Pool? Explained," *The Hindu*, December 11, 2023.
18. Satish Deshpande, "Rethinking Majority and Minority in India: The Question of Caste Census" (Counting Caste: Breaking the Caste Census Deadlock, University of Oxford, February 5, 2022).
19. Vivek Kumar Singh, "Counting Caste: A Step towards Radical Anticasteism." *Prabuddha: Journal of Social Equality* 7, no. 1 (2023): 8–9.
20. Ibid., 1.
21. Ankur Paliwal, "How India's Caste System Limits Diversity in Science—in Six Charts," *Nature* 613, no. 7943 (2023): 230–34; Murali Shanmugavelan, "May the Myth of Castelessness Die," *LOGIC(S)*, no. 19 (2023); Ashok Danavath, "A Discriminatory Education Policy That Further Excludes the Oppressed from Academia: The Case of the National Overseas Scholarship (NOS) for SC-ST Scholars in India" (Rapoport Center Human Rights Working Paper Series, The University of Texas at Austin, August 2023); Palashi Vaghela, Steven J. Jackson, and Phoebe Sengers, "Interrupting Merit, Subverting Legibility: Navigating Caste in "Casteless" Worlds of Computing," in *Proceedings of the 2022 CHI Conference on Human Factors in Computing Systems (CHI "22)*, article 545 (New York, NY: Association for Computing Machinery, 2022), 1–20; P. Kureel, "Indian Media and

Caste," *Caste: A Journal on Social Exclusion* 2, no. 1 (2021): 97–108; Subramanian, *The Caste of Merit*; Subramanian, "Making Merit"; Robin Jeffrey, "Missing from the Indian Newsroom," *The Hindu*, April 9, 2012; Surinder Jodhka and Katherine Newman, "In the Name of Globalisation: Meritocracy, Productivity and the Hidden Language of Caste," *Economic and Political Weekly* 42, no. 41 (October 13, 2007): 4125–32; S. Anand, "Covering Caste: Visible Dalit, Invisible Brahmin," in *Practising Journalism: Values, Constraints, Implications*, ed. N. Raj, 172–97 (New Delhi: Sage, 2005).

Appendices

Appendix A: Socio-Economic Caste (SEC) Survey Questionnaire (Urban)[1]

For Each Household

Section 1: Housing/Dwelling

1. Predominant material of wall of the dwelling room (s) (Give code)
   1 = Grass/thatch/bamboo, etc.
   2 = Plastic/polythene
   3 = Mud/unburnt brick
   4 = Wood
   5 = Stone not packed with mortar
   6 = Stone packed with mortar
   7 = G.I./metal/asbestos sheets
   8 = Burnt brick
   9 = Concrete
   0 = Any other
2. Predominant material of roof of the dwelling room (s) (Give code)
   1 = Grass/thatch/bamboo/wood/mud, etc.
   2 = Plastic/polythene
   3 = Hand-made tiles
   4 = Machine-made tile
   5 = Burnt brick
   6 = Stone
   7 = Slate
   8 = G.I./metal/asbestos sheets

9 = Concrete
0 = Any other

3. Ownership status of this house (Give code)
   1 = Owned
   2 = Rented
   3 = Shared
   4 = Living on premises with employer
   5 = House provided by employer
   6 = Any other
4. Number of dwelling rooms exclusively in possession of this household (Record 1,2,3…)

Section 2: Amenities

5. Availability of drinking water source
   1 = Within the premises, 2 = Near the premises, 3 = Away
6. Main source of lighting
   1 = Electricity, 2 = Kerosene, 3 = Solar, 4 = Other oil, 5 = Any other, 6 = No lighting
7. Water-seal latrine exclusively for the household
   1 = Yes, 2 = No
8. Wastewater outlet connected to
   1 = Closed drainage, 2 = Open drainage, 3 = No drainage
9. Separate room used as kitchen exclusively for the household
   1 = Yes, 2 = No

Section 3: Assets

Does the household own the following assets (Give code)

10. Refrigerator
    1 = Yes; 2 = No
11. Telephone/Mobile phone
    Yes: 1 = Landline only, 2 = Mobile only, 3 = Both, 4 = No
12. Computer/Laptop
    Yes: 1 = With internet, 2 = Without internet, 3 = No
13. Motorized wheelers
    1 = Two/Three-wheeler, 2 = Four-wheeler, 3 = No
14. AC
    1 = Yes; 2 = No
15. Washing machine
    1 = Yes; 2 = No

For Each Person

1. Serial No.
2. Name of the person (start with the head of the household)
3. Relationship to head (record the relationship in full)
4. Sex
   1 = Male
   2 = Female
   3 = Transgender[2]
5. Date of birth (as per English calendar)
6. Marital status
   1 = Never married
   2 = Currently married
   3 = Widowed
   4 = Separated
   5 = Divorced
7. Name of father
8. Name of mother
9. Occupation/Activity (describe the actual work)
10. Highest educational level completed
    1 = Illiterate
    2 = Literate but below primary
    3 = Primary
    4 = Middle
    5 = Secondary
    6 = Higher secondary
    7 = Graduate or higher
    8 = Other (Specify)
11. Main source of income/earnings from work
    1 = Beggar/rag-picker
    2 = Domestic worker
    3 = Street vendor/cobbler/hawker/other service provider working on streets
    4 = Construction worker/plumber/mason/labor/painter/welder/security guard/coolie and other head-load worker
    5 = Sweeper/sanitation worker/mali
    6 = Home-based worker/artisan/handicrafts worker/tailor
    7 = Transport worker/driver/conductor/helper to drivers and conductors/cart puller/rickshaw puller
    8 = Shop worker/assistant/peon in small establishment/helper/delivery assistant/attendant/waiter

9 = Electrician/mechanic/assembler/repair worker
10 = Washer-man/chowkidar
11 = Other work/Non-work
12 = Non-work (Pension/Rent/Interest, etc.)

12. Are wages earned (Give code)
    1 = daily
    2 = weekly
    3 = monthly
    4 = irregularly
    5 = not wage earner
13. Disability
    1 = In seeing
    2 = In hearing
    3 = In speech
    4 = In movement
    5 = Mental retardation
    6 = Mental illness
    7 = Other disability
    8 = Multiple disability
    9 = Not disabled
14. Chronic illness
    1 = Cancer
    2 = HIV/AIDS
    3 = TB
    4 = Leprosy
    5 = Other illness
    6 = No chronic illness
15. Religion (write name of the religion in full)
16. Caste/Tribe Status (Give code):
    Scheduled Caste (SC)-1
    Scheduled Tribe (ST)-2
    Other-3
    No Caste/Tribe-4
    (SC can be only among the Hindus, Sikhs, and Buddhists. ST can be from any religion.)
17. If code 1, 2, or 3 in Col. 16, write name of Caste/Tribe. If code 4 in Col. 16, put "X."

Appendix B: Research Methodology and Support

I used an extended case study methodology to explore the politics surrounding the counting of caste in Census 2011 and the practices of enumerating caste in an alternate state project. The case study methodology examines a historical episode, or an aspect of a historical episode, to develop or test explanations that may be generalizable to other events.[3] The extended case study methodology, which is an extension of the case study, sets out to observe the world from the participants' standpoint without losing the ability to explain intricacies in more generalizable terms and while being reflexive about the researcher's own role.[4] As an Indian-American with family origins in the south Indian state of Kerala, and in the Syrian Christian community in particular, I have been drawn to making sense of how inequality works in the United States and India (each serving a 'shadow case" for the other). Syrian Christians have a high position in the social hierarchy in Kerala and considerable political and economic power within the state. As a sociologist, I was drawn to examining a historical event that suddenly made visible the messiness of social-knowledge-making processes and the politics of caste. Sociologists Charles Camic, Neil Gross, and Michele Lamont argue that understanding social knowledge-making requires studying concrete social locations that are "relatively porous," such that "factors that sociologists ordinarily associate with different levels [for example, macro and micro] combine."[5] Much of my initial fieldwork involved observations of data collection operations and interviews with individuals responsible for overseeing these operations in concrete locations, from local data-processing centers to state and centralized offices overseeing data collection in the south Indian state of Karnataka, where I had previously lived and worked. As I struggled to make sense of why the central government did not collect the caste-wise data in Census 2011, after it had agreed to do so, and why the caste census became part of a state project that had previously received so much criticism, and the messiness of that data collection process that I was observing unfold, I ventured more into meso- and macro-level knowledge-making spaces. These conversations helped me to interpret my observations of household interviews. During the process of synthesizing and analyzing my own observational and interview data, I read the works of anti-caste scholar-activists and came to understand how the SEC survey is a highly visible case of how castelessness is contested and reveals several contemporary mechanisms of Brahmanism.

I collected observational, interview, and document data to study the case. I found that approaching the caste census from multiple vantage points was helpful to triangulate what was happening and to examine how a variety of context-specific actors and processes interact, unfold, and have consequence. For example,

I learned that information about the challenges that emerged during household interviews and processes of enumeration may be omitted by senior bureaucrats (both because it takes time for this data to filter back up through the bureaucratic hierarchy and because they would prefer not to discuss these details with a private researcher) but freely discussed in conversations with government workers in the lower levels of the bureaucracy and observed while accompanying data collection teams. I use pseudonyms for individuals discussed in my field notes.

I conducted fieldwork during seven trips to India, ranging from three weeks to two months in length between July 2011 and August 2016, for a total of nine months. My data collection began at the start of field operations for the SEC survey in July 2011 and across several trips that continued through the period of re-enumeration in select areas in early 2013. I returned to India in 2015 and 2016 for follow-up interviews with key informants. I resorted to this unusual pattern of data collection during my Ph.D. research because I had a young child (18 months old) when I began fieldwork in July 2011 and gave birth to a second child in mid-2013. In addition, while the Indian government initially planned and publicized that the SEC survey would take six months to complete, the project ended up spanning several years to finish.[6]

The SEC survey involved a door-to-door enumeration of every household in the country, and I wished to observe this process. My first research trip was in July and August 2011, as SEC survey operations were beginning in select regions of the country. During this trip, I visited Bengaluru, Thiruvananthapuram, and Delhi to understand the regional setup for SEC survey operations and the centralized management of the SEC survey in the nation's capital. In this trip and my subsequent research trips, I interviewed and talked to government officials, including senior bureaucrats, consultants, and lower-level civil servants, public intellectuals, civil society leaders, private sector employees involved in data entry, and private citizens. In total, for this project, I conducted about ninety semi-structured interviews and a research assistant completed 29 additional interviews (see Table A.1). During my interviews, I took detailed notes but did not use a recording device, as doing so would likely have restricted the types of questions that I could ask and the types of answers that respondents would provide. Therefore, all my material is a close approximation of the interview, not a direct transcription.

I made two subsequent research trips when the main SEC survey operations were occurring in late 2011 and early 2012. During the first trip in December 2011, I observed the setup of data-processing centers, participated in enumerator trainings, and accompanied data collection teams while they interviewed households in two enumeration blocks in a ward in the periphery of Bengaluru. I returned to Bengaluru in February and March of 2012 when survey operations were in full swing and observed household interviews in four enumeration blocks

Table A.1 Interviews Conducted during Fieldwork

		Interviews conducted
1	**Public sector (current and former)**	**25**
	A. Central government	4
	B. State government	6
	C. Municipal government	15
2	**Civil society organizations**	**25**
	A. National-level organizations	1
	B. Local organizations (caste associations)	12
	C. Other members of civil society (scholars, journalists)	12
3	**Private sector**	**11**
	A. Managers	3
	B. Workers	8
4	**Private enumerated households/individuals***	**29**
	A. Bangalore Urban District	25
	B. Bangalore Rural District	4
	Total	**90**

Note: *A Bengaluru-based research assistant interviewed 29 enumerated households.

in a centrally located ward. Across these two trips, I observed approximately 250 household interviews in six different neighborhoods in two wards in Bangalore. After observing about 20 household interviews in December 2011, I designed a semi-structured protocol to interview families after they had been enumerated for the SEC survey. A Bengaluru-based research assistant, Shalini Jamuna Kotresh, conducted 29 interviews with enumerated families between January and May 2012, while I continued observations of the SEC survey. She purposively selected households across religious, caste, and socioeconomic lines in Bangalore Urban and Rural districts.

I made two additional research trips in mid-2012 and early 2013, after the main survey operations in Bengaluru began to wind down but before the SEC survey data were finalized. In July 2012, I returned to Bengaluru to observe data processing. During this trip, I visited neighborhood data-processing centers and interviewed senior government officials involved in survey operations in Karnataka and caste association staff. In my conversations with individuals connected with caste association organizations, I learned about the general history and practices of the organizations and if and how they tried to mold public perceptions, participation, and responses in relation to the enumeration of caste. I also collected caste association magazines and newsletters for the period surrounding the survey, when available. In January and February 2013, I observed the re-enumeration of 100 households in a fourth neighborhood in Bengaluru. Re-enumeration was extremely limited; for example, in Bangalore's East Zone (the city is split into three geographic zones), only 5 out of 3,241 enumeration blocks were re-enumerated. My fieldwork across these first five research trips was ethnographic in its approach, as I tried to understand the day-to-day processes of collecting and producing caste data through observations of survey operations and interviews of survey officials and data collection teams.

My final two research trips occurred after the completion of the SEC survey ground operations and soon after the government published the SEC data—except for the caste-wise data—which remain unpublished. During these trips in the summer of 2015 and 2016, I sought to better understand the chain of events that unfolded between late 2009 and early 2011, when caste census advocates organized to change census policy on caste and appeared to succeed after securing a public concession from top political leaders, but then the executive bureaucracy maneuvered to push the caste count into an alternative state project. These interviews took on new meaning after I had observed SEC survey operations, and I sought to make sense of why the eventual caste survey bore little resemblance to the demands made by caste census advocates.

The primary documents related to this project come from three main sources. The first set of documents includes government-produced materials connected to the SEC survey. This includes training manuals; survey schedules; the data entry program for the SEC survey; government circulars, letters, and memoranda; and additional survey documents. The second source of documents comes from the media: English, Kannada, Tamil, and Hindi newspapers. Three India-based research assistants catalogued all caste-related and SEC-survey-related articles for a one-year period surrounding the survey. Shalini Jamuna Kotresh, who also interviewed 29 enumerated families in Karnataka, created an archive of the *Prajavaani* (Kannada) paper; Arasi Arivu catalogued articles in the *Dhinathanthi* (Tamil) newspaper from the neighboring state of Tamil Nadu; Devanshu Singh analyzed the *Dainik Jagran* (Hindi), which is read in the Hindi-speaking regions of north India. I created a similar archive for several Indian English newspapers: *The Hindu, Times of India, Indian Express, Hindustan Times, and Deccan Herald.*

While the collected SEC survey documents helped to provide a centralized view of survey operations, the newspaper coverage offered a comparative perspective on how the SEC survey operations and publicity unfolded in different regions of the country. The third set of documents focuses on historical censuses—the questionnaires, data collection plans, and printed reports from colonial censuses. These documents provided a crucial historical view of caste in the decennial census. The large data collection team involved in this project was central to completing this manuscript.

Student research assistants have helped to collect secondary documents for this project and related research: Natasha Allen, Molly Andrus, Kellen Buckley, Christina Charie, Yingting (Jenny) Chen, Savannah Dhar, Kathryn Doner, Nicole Durant, Grace Feisthamel, Dylan Flaherty, Briana Gittens, Lisa McNamara, Tatianna Medina, Lela Miller, Diana Mora, Dennis Mueñtes, Angie Pierre, Alicia Terrero, Sydaya Tompkins, Ana Martinez Vargas, and Jennifer Vargas. The large data collection team involved in compiling and analyzing primary and secondary data was central to completing this manuscript and I'm extremely grateful for the collective effort that made this book possible.

Notes

1. The SEC survey rural questionnaire included questions on the dwelling structure (Section 1) and household assets (Section 4) that were similar to the urban questionnaire but not identical. The rural questionnaire had three additional sections focused on the household: (*a*) three questions about whether any member of the household was from a primitive tribal group, a legally released bonded laborer, or a manual scavenger (Section 2); (*b*) six questions related to employment and income (Section 3); (*c*) four questions related to land ownership and three questions related to household assets (that is, mechanized three/four wheel agricultural equipment, irrigation equipment, and a Kisan credit card with a credit limit of 50,000 INR or above) (Section 5). The individual-level questions in the rural questionnaire did not include questions on main source of income or frequency of wages (although some of this data was captured at the household level in Section 3).
2. After the SEC survey started, the question on sex was updated include an additional answer option of transgender.
3. Alexander George and Andrew Bennett, *Case Studies and Theory Development in the Social Sciences* (Cambridge, MA: MIT Press, 2005).
4. Michael Burawoy, "The Extended Case Method" *Sociological Theory* 16, no. 1 (1998): 4–33.
5. Charles Camic, Neil Gross, and Michele Lamont, *Social Knowledge in the Making* (Chicago: Chicago University Press, 2011), 1.
6. Generous funding made multiple research trips possible. I detail my sources of funding in the acknowledgements.

References

Ahmad, Imtiaz. *Caste and Social Stratification among the Muslims*. Delhi: Manohar Book Service, 1973.

———. "The Ashraf–Ajlaf Dichotomy in Muslim Social Structure in India." *Indian Economic and Social History Review* 3, no. 3 (1966): 268–78.

Ahmad, Imtiaz, and Shashi Bhushan Upadhyay, eds. *Dalit Assertion in Society, Literature and History*. Delhi: Orient BlackSwan, 2010.

Ahmad, Irfan. "A Different Jihad: Dalit Muslims' Challenge to Ashraf Hegemony." *Economic and Political Weekly* 38, no. 46 (2003): 4886–91.

Alam, Anwar. "Democratisation of Indian Muslims: Some Reflections." *Economic and Political Weekly* 38, no. 46 (November 15, 2003): 4881–85.

Alexander, K. C. "The Problem of Caste in the Christian Churches of Kerala." In *Caste among Non-Hindus in India*, edited by Harjinder Singh, 50–65. Delhi: National Publishing House, 1977.

Alkire, Sabina, and Suman Seth. "Determining BPL Status Some Methodological Improvements." *Indian Journal of Human Development* 2, no. 2 (2008): 407–24.

———. "Identifying BPL Households: A Comparison of Methods." *Economic and Political Weekly* 48, no. 2 (January 12, 2013): 49–57.

———. "Selecting a Targeting Method to Identify BPL Households in India." *Social Indicators Research* 112, no. 2 (2013): 417–46.

Allendorf, Keera, and Roshan Pandian. "The Decline of Arranged Marriage? Marital Change and Continuity in India." *Population and Development Review* 42, no. 3 (2016): 435–64.

Aloysius, G. *Nationalism without a Nation in India*. Delhi: Oxford University Press, 1997.

———. *Religion as Emancipatory Identity: A Buddhist Movement among the Tamils under Colonialism*. New Delhi: New Age International, 1998.

Ambedkar, B. R. "Annihilation of Caste." In *Dr. Babasaheb Ambedkar: Writings and Speeches*, edited by Vasant Moon, Vol. I, 23–98. New Delhi: Dr. Ambedkar

Foundation, Ministry of Social Justice and Empowerment, Government of India, 2014. Accessed April 1, 2024. https://www.mea.gov.in/Images/attach/amb/Volume_01.pdf.

———. "Letter from B. R. Ambedkar to W. E. B. Du Bois." Undated (likely written in 1946). South Asian American Digital Archive. Accessed April 1, 2024. https://www.saada.org/item/20140415-3544.

———. "Statement by Dr. B. R. Ambedkar in Parliament in Explanation of His Resignation from the Cabinet." In *Dr. Babasaheb Ambedkar: Writings and Speeches*, edited by Vasant Moon, Vol. XIV (2), 1315–27. New Delhi: Dr. Ambedkar Foundation, Ministry of Social Justice and Empowerment, Government of India, 2013. Accessed April 1, 2024. https://www.mea.gov.in/Images/attach/amb/Volume_14_02.pdf.

———. "States and Minorities." In *Dr. Babasaheb Ambedkar: Writings and Speeches*, edited by Vasant Moon, Vol. I, 381–449. New Delhi: Dr. Ambedkar Foundation, Ministry of Social Justice and Empowerment, Government of India, 2014.

———. "The Buddha and His Dhamma." In *Dr. Babasaheb Ambedkar: Writings and Speeches*, edited by Vasant Moon, Vol. XI, 9–599. New Delhi: Dr. Ambedkar Foundation, 1979.

———. "Thoughts on Linguistic States." In *Dr. Babasaheb Ambedkar: Writings and Speeches*, edited by Vasant Moon, Vol. I, 139–70. New Delhi: Dr. Ambedkar Foundation, Ministry of Social Justice and Empowerment, Government of India, 2014. Accessed April 1, 2024. https://www.mea.gov.in/Images/attach/amb/Volume_01.pdf.

———. "Who Are the Shudras? How They Came to Be the Fourth Varna in Indo-Aryan Society (1947)." In *Dr. Babasaheb Ambedkar: Writings and Speeches*, edited by Vasant Moon, 3rd ed., Vol. VII, 5–227. New Delhi: Dr. Ambedkar Foundation, Ministry of Social Justice and Empowerment, Government of India, 2020.

Amenta, Edwin, Neal Caren, Elizabeth Chiarello, and Yang Su. "The Political Consequences of Social Movements." *Annual Review of Sociology* 36, no. 1 (2010): 287–307.

Anand, S. "Covering Caste: Visible Dalit, Invisible Brahmin." In *Practising Journalism: Values, Constraints, Implications*, edited by N. Raj, 172–97. New Delhi: Sage, 2005.

Anderson, Benedict. *Imagined Communities: Reflections on the Origin and Spread of Nationalism*. London: Verso, 1983.

Andrews, Kenneth T. "Social Movements and Policy Implementation: The Mississippi Civil Rights Movement and the War on Poverty, 1965 to 1971." *American Sociological Review* 66, no. 1 (2001): 71–95.

Ansari, Ghaus. *Muslim Caste in Uttar Pradesh: A Study in Culture Contact.* Lucknow: Ethnographic and Folk Culture Society, 1960.

Ansari, Khalid Anis. "India's Muslim Community under a Churn: 85% Backward Pasmandas up against 15% Ashrafs." *The Print*, May 13, 2019, sec. Opinion. Accessed April 1, 2024. https://theprint.in/opinion/indias-muslim-community-under-a-churn-85-backward-pasmandas-up-against-15-ashrafs/234599/.

Appadurai, Arjun. "Number in the Colonial Imagination." In *Orientalism and the Postcolonial Predicament: Perspectives on South Asia*, edited by Carol A. Breckenridge and Peter van der Veer, 314–36. South Asia Seminar Series. Philadelphia: University of Pennsylvania Press, 1993.

Arya, Sunaina. "Dalit or Brahmanical Patriarchy? Rethinking Indian Feminism." *CASTE/A Global Journal on Social Exclusion* 1, no. 1 (2020): 217–28.

Arya, Sunaina, and Aakash Singh Rathore, eds. *Dalit Feminist Theory: A Reader.* London: Routledge, 2019.

Asian Age. "Longer Wait for Caste Census Information." January 30, 2017. Accessed April 1, 2024. https://www.asianage.com/india/all-india/300117/longer-wait-for-caste-census-information.html.

Ayyathurai, Gajendran. "Colonialism, Caste, and Gender: The Emergence of Critical Caste Feminism in Modern South India." *Journal of Women's History* 33, no. 3 (Fall 2021): 133–56.

Bachchan, Amitabh. "Day 750." *Bachchan Bol* (blog), May 11, 2010. Accessed April 1, 2024. http://srbachchan.tumblr.com/post/25143096756.

Backward Classes Commission (B.P. Mandal, Chair). *Report of the Backward Classes Commission.* Delhi: Government of India, 1980. https://ruralindiaonline.org/en/library/resource/report-of-the-backward-classes-commission-volumes-i-and-ii/#section02.

Backward Classes Commission (K. Kalelkar, Chair). *Report of the Backward Classes Commission.* Simla: Government of India, 1955.

Baines, J. A. "Conference of Census Officers of the 1881 Census to Consider the Matter Relating to 1891 Census (Feb 10th, 1890)." In *Indian Census through a Hundred Years, Census of India 1971*, by Dandapani Natarajan, 243–47. Delhi: Office of the Registrar General, India, 1972.

———. "Letter to George Hamilton (April 29, 1899)." In *Indian Census through a Hundred Years, Census of India 1971*, by Dandapani Natarajan, 247–52. Census of India 1971. Delhi: Office of the Registrar General, India, 1972.

Baiocchi, Gianpaolo, and Ernesto Ganuza. *Popular Democracy: The Paradox of Participation.* Stanford University Press, 2020.

Bala, J. "Pandit Iyothee Thass and the Revival of Tamil Buddhism." Round Table India for An Informed Ambedkar Age, June 27, 2014. Accessed April 1, 2024.

https://www.roundtableindia.co.in/pandit-iyothee-thass-and-the-revival-of-tamil-buddhism/.

Basheer, K. P. M. "Nature of Caste Census Still Unclear." *The Hindu*, December 13, 2010.

Bayly, Susan. *Caste, Society and Politics in India from the Eighteenth Century to the Modern Age*. The New Cambridge History of India. Cambridge: Cambridge University Press, 1999.

BBC News. "India Approves Caste Based Census." August 12, 2010.

Beteille, Andre. "The Reproduction of Inequality: Occupation, Caste and Family." *Contributions to Indian Sociology* 25, no. 1 (1991): 3–28.

Blum, Alain. "Resistance to Identity Categorization in France." In *Census and Identity: The Politics of Race, Ethnicity, and Language in National Censuses*, edited by David I. Kertzer and Dominique Arel, 121–47. Cambridge: Cambridge University Press, 2002.

Bourdieu, Pierre. *Language and Symbolic Power*. Edited by John Thompson. Cambridge: Polity Press, 1991.

———. *Practical Reason: On the Theory of Action*. Stanford, CA: Stanford University Press, 1998.

Bowker, Geoffrey C. *Science on the Run: Information Management and Industrial Geophysics at Schlumberger, 1920–1940*. Cambridge, MA: MIT Press, 1994.

Brass, Paul. "The Politicization of the Peasantry in a North Indian State." *The Journal of Peasant Studies* 7, no. 4 (n.d.): 395–426.

Brown, Mark B. *Science in Democracy: Expertise, Institutions, and Representation*. Cambridge, MA: MIT Press, 2009.

Burawoy, Michael. "The Extended Case Method." *Sociological Theory* 16, no. 1 (1998): 4–33.

Camic, Charles, Neil Gross, and Michele Lamont. *Social Knowledge in the Making*. Chicago: University of Chicago Press, 2011.

Centeno, Miguel Angel. *Democracy within Reason: Technocratic Revolution in Mexico*. University Park, PA: Penn State Press, 2010.

Chakravarti, Uma. "Conceptualizing Brahmanical Patriarchy in Early India: Gender, Caste, Class, and State." *Economic and Political Weekly* 28, no. 14 (1993): 579–85.

———. *Gendering Caste through a Feminist Lens*. Delhi: Sage, 2018.

———, ed. *Thinking Gender Doing Gender: Feminist Scholarship and Practice Today*. Hyderabad: Orient BlackSwan, 2016.

Chauhan, Chetan. "The Importance of Caste Census and Its Political Implications." *Hindustan Times*, February 26, 2021. Accessed April 1, 2024. https://www.hindustantimes.com/india-news/importance-of-caste-census-and-its-political-implications-101614336418488.html.

———. "Updated OBC Count: 41 Pc." *Hindustan Times*, November 1, 2006. Accessed April 1, 2024. https://www.hindustantimes.com/india/updated-obc-count-41-pc/story-uyhXtHE6P4bg0veGknfpkN.html.

Chesney, George. *Indian Polity: A View of the System of Administration in India*. London: Longmans, Green and Co., 1894.

Chibber, Maneesh. "GoM Unanimous Yes to Caste in Census." *Indian Express*, August 12, 2010.

Chief Economic Adviser, Government of India. "Document No. H-1102/8/2010-CD(CEN)." Ministry of Rural Development, October 26, 2012.

Chowdhury, Sohini. "Supreme Court Refuses to Interfere with High Court's Stay on Bihar Caste Survey as HC Is Set to Hear Main Matter in July." *Live Law*, May 18, 2023.

Cohn, Bernard. *Colonialism and Its Forms of Knowledge*. Princeton, NJ: Princeton University Press, 1996.

———. "The Census, Social Structure and Objectification in South Asia." In *An Anthropologist among the Historians and Other Essays*, 224–54. Delhi: Oxford University Press, 1987.

Collins, Harry, and Robert Evans. *Rethinking Expertise*. Chicago: University of Chicago Press, 2007. https://doi.org/10.7208/9780226113623.

Consortium of Public Sector Undertakings (CPSUs) Tender Processing PSU ITI Ltd. "Socio-Economic Caste Census-2011, Request for Quotation: Service Contract for Data Collection on Door to Door Basis for 120 Crore Population," July 18, 2011.

"Constituent Assembly Debates, Official Report." New Delhi: Lok Sabha Secretariat, 1949.

Constitution of India as Amended by Constitution Act 2009. Delhi: Kamal Publishers, 2011.

Curtis, Bruce. *The Politics of Population: State Formation, Statistics and the Census of Canada 1840–1875*. Toronto: University of Toronto Press, 2001.

Dalit Panthers. *Manifesto*. Bombay, 1973. Accessed April 1, 2024. http://www.raiot.in/dalit-panthers-manifesto/.

Danavath, Ashok. "A Discriminatory Education Policy That Further Excludes the Oppressed from Academia: The Case of the National Overseas Scholarship (NOS) for SC–ST Scholars in India." Rapoport Center Human Rights Working Paper Series, The University of Texas at Austin, August 2023.

———. "Telangana's Backward Caste Overseas Scholarship Hampered by Inclusion of EWS Reservation Policy." *News Click*, April 27, 2022. Accessed April 1, 2024. https://www.newsclick.in/Telangana-Backward-Caste-Overseas-Scholarship-Hampered-Inclusion-EWS-Reservation-Policy.

Danavath, Ashok, and Jyoti Bania. "Why Aren't Enough Students from Tribal Communities Receiving the National Overseas Scholarship?" *The Wire*, July 27, 2021. Accessed April 1, 2024. https://thewire.in/education/govt-overseas-scholarship-criteria-sc-st-students.

Deccan Herald. "Caste Census Data Riddled with Errors." July 28, 2015. Accessed April 1, 2024. https://www.deccanherald.com/content/492081/caste-census-data-riddled-errors.html.

———. "Mukherjee to Head GoM Caste-Based Census Panel." June 9, 2010.

Desai, Sonalde. "Reimagining the OBC Quota." *The Hindu*, September 19, 2017, sec. Opinion. Accessed April 1, 2024. https://www.thehindu.com/opinion/op-ed/reimagining-the-obc-quota/article19710515.ece.

Deshpande, G. P., "Introduction." In *Selected Writings of Jotirao Phule*, edited by G. P. Deshpande, 1–21. New Delhi: LeftWord Books, 2002.

———. "On Infant Marriage and Enforced Widowhood." In *Selected Writings of Jotirao Phule,* edited by G. P. Deshpande, 191–92. New Delhi: LeftWord Books, 2002.

Deshpande, Satish. "Caste and Castelessness: Towards a Biography of the General Category." *Economic and Political Weekly* 48, no. 15 (2013): 32–39.

———. *Contemporary India: A Sociological View*. Delhi: PenguinViking, 2003.

———. "Rethinking Majority and Minority in India: The Question of Caste Census." Presentation at the International Conference on Counting Caste: Breaking the Caste Census Deadlock, University of Oxford, February 5, 2022.

Deshpande, Satish, and Geetika Bapna. *Dalits in the Muslim and Christian Communities*. New Delhi: National Commission for Minorities, Government of India, 2008.

Desrosières, Alain. *The Politics of Large Numbers: A History of Statistical Reasoning*. Cambridge, MA: Harvard University Press, 1998.

Dhar, Aarti. "Caste in Census: Cabinet to Decide on Modalities." *The Hindu*, August 12, 2010.

———. "Now, Caste Census Is Just a Formality, Says Centre." *The Hindu*, September 1, 2010.

Dirks, Nicholas. *Castes of Mind: Colonialism and the Making of Modern India*. Princeton, NJ: Princeton University Press, 2001.

———. "Foreword." In *Colonialism and Its Forms of Knowledge*, by Bernard Cohn, ix–xvii. Princeton, NJ: Princeton University Press, 1996.

Doner, Kathryn Victoria Bahnken. "Seventy Years Later: Caste in the Indian Bureaucracy." *Sociology Between the Gaps: Forgotten and Neglected Topics* 7 (2022). Accessed April 1, 2024. https://digitalcommons.providence.edu/sbg/vol7/iss1/7.

Drèze, Jean. "The Revolt of the Upper Castes." *CASTE/A Global Journal on Social Exclusion* 1, no. 1 (2020): 229–36.

Drèze, Jean, and Reetika Khera. "The BPL Census and a Possible Alternative." *Economic and Political Weekly* 45, no. 9 (February 27, 2010): 54–63.

Dutt, Barkha. "In Reverse Gear." *Hindustan Times*, May 14, 2010.

Economic Times. "Government Defends Deferment of Caste Data; 8.19 Crore Errors Found." July 28, 2015. Accessed April 1, 2024. https://economictimes.indiatimes.com/news/politics-and-nation/government-defends-deferment-of-caste-data-8-19-crore-errors-found/articleshow/48255008.cms?from=mdr.

———. "No Plan to Release 2011 Caste Census Data: Govt." July 26, 2022. Accessed April 1, 2024. https://economictimes.indiatimes.com/news/india/no-plan-to-release-2011-caste-census-data-govt/articleshow/93139528.cms.

Espeland, Wendy Nelson, and Mitchell L. Stevens. "A Sociology of Quantification." *European Journal of Sociology* 49, no. 3 (2008): 401–36.

———. "Commensuration as a Social Process." *Annual Review of Sociology* 24 (1998): 313–43.

Financial Express. "In a First, Government to Collect Data on OBCs in Census 2021." September 1, 2018. Accessed April 1, 2024. https://www.financialexpress.com/india-news/in-a-first-government-to-collect-data-on-obcs-in-census-2021/1299461.

Foucault, Michel. *Discipline and Punish*. New York: Vintage Books, 1979.

———. "Governmentality." In *The Foucault Effect: Studies in Governmentality*, edited by Graham Burchell, Colin Gordon, and Peter Miller, 87–104. Chicago: University of Chicago Press, 1991.

———. "Power and Strategies." In *Power/Knowledge: Selected Interviews and Other Writings, 1972–1977*, edited by Colin Gordon, 134–45. New York: Pantheon Books, 1980.

Galanter, Marc. *Competing Equalities*. Berkeley: University of California Press, 1984.

Gandhi, M. K. "Harijan (5 January 1934)." In *Collected Works of Mahatma Gandhi (8 October 1933–17 January 1934)*, Vol. 62. New Delhi: Publications Division Government of India, 1999. Accessed April 1, 2024. http://www.gandhiashramsevagram.org/gandhi-literature/mahatma-gandhi-collected-works-volume-62.pdf.

———. "India Round Table Conference (Second Session): Proceedings of Federal Structure Committee and Minorities Committee." In *Collected Works of Mahatma Gandhi (13 October 1931–8 February 1932)*, Vol. 54. New Delhi: Publications Division Government of India, 1999. Accessed April 1, 2024.

http://www.gandhiashramsevagram.org/gandhi-literature/mahatma-gandhi-collected-works-volume-54.pdf.

Geetha, V., and S. V. Rajadurai. "Dalits and Non-Brahman Consciousness in Colonial Tamil Nadu." *Economic and Political Weekly* 28, no. 39 (September 25, 1993): 2091–98.

———. *Towards a Non-Brahman Millennium from Iyothee Thass to Periyar.* Calcutta: Samya, 1998.

George, Alexander and Andrew Bennett. *Case Studies and Theory Development in the Social Sciences*. Cambridge, MA: MIT Press, 2005.

Ghildiyal, Subodh. "Is It End of Road for First Caste Census?" *Times of India*, October 10, 2017. Accessed April 1, 2024. https://timesofindia.indiatimes.com/india/is-it-end-of-road-for-first-caste-census/articleshow/61013698.cms.

Ghurye, G. S. *Caste and Race in India*. London: K. Paul, Trench, Trubner & Company, Limited, 1932.

Gibson, Christopher L. "The Consequences of Movement Office-Holding for Health Policy Implementation and Social Development in Urban Brazil." *Social Forces* 96, no. 2 (2017): 751–78.

Giugni, Marco G. "Was It Worth the Effort? The Outcomes and Consequences of Social Movements." *Annual Review of Sociology* 24 (1988): 371–93.

Gool, Sebastiaan Maria van. "Untouchable Bureaucracy: Unrepresentative Bureaucracy in a North Indian State." Doctoral thesis, Leiden University, 2008. Accessed April 1, 2024. https://hdl.handle.net/1887/12631.

Government of India. "Socio-Economic Survey 2010: Supervisor Manual." Ministry of Rural Development, New Delhi, 2010.

Government of India (Cabinet), Press Information Bureau. "Enumeration of Castes Other than Scheduled Castes and Scheduled Tribes." News release, September 9, 2010. Accessed April 1, 2024. https://pib.gov.in/newsite/PrintRelease.aspx?relid=65631.

Government of India (Ministry of Finance), Press Information Bureau. "Provisional Data of Socio Economic and Caste Census (SECC) 2011 for Rural India Released." News release, July 3, 2015. Accessed April 1, 2024. https://pib.gov.in/newsite/PrintRelease.aspx?relid=122963.

Government of India (Ministry of Rural Development), Press Information Bureau. "Below Poverty Line (BPL) Census to Be Conducted along with Caste Census during June-December 2011." News release, May 19, 2011. Accessed April 1, 2024. https://pib.gov.in/newsite/PrintRelease.aspx?relid=72217.

———. "Current BPL Census to Rectify Past Mistakes." News release, August 11, 2011. Accessed April 1, 2024. https://archive.pib.gov.in/newsite/PrintRelease.aspx?relid=74356.

Government of India (N. C. Saxena, Chair). *Report of the Expert Group to Advise the Ministry of Rural Development on the Methodology for Conducting the Below Poverty Line (BPL) Census for the 11th Five-Year Plan*. New Delhi: Government of India, Ministry of Rural Development, August 2009. Accessed April 1, 2024. https://rural.nic.in/sites/default/files/ReportofExpertGroupChaired-Dr.N.C.Saxena.pdf.

Government of Mysore. *Report of the Sir Leslie Miller Committee*. 1919.

Government of Mysore (Nagana Gowda, Chair). *Mysore Backward Classes Committee Final Report*. Bangalore: Government Press, 1961.

Guha, Sumit. "The Politics of Identity and Enumeration in India c. 1600–1990." *Comparative Studies in Society and History* 45, no. 1 (2003): 148–67.

Gupta, Akhil. *Red Tape: Bureaucracy, Structural Violence, and Poverty in India*. Durham: Duke University Press, 2012.

Gupta, Dipankar. "Caste and Politics: Identity over System." *Annu. Rev. Anthropol.* 34 (2005): 409–27.

Gupta, Smita. "Centre Clears Caste in Census 2010." *The Hindu*, September 9, 2010.

———. "Deadline Nears for Parties to Take Stand on Caste Census." *The Hindu*, August 2, 2010.

———. "GoM Approval for Caste-Based Census." *The Hindu*, August 11, 2010.

Gupta, Smita, and Aarti Dhar. "Stand-Alone Caste Headcount Too a Certainty." *The Hindu*, August 31, 2010.

Guru, Gopal. "Dalit Women Talk Differently." *Economic and Political Weekly* 30, no. 41–42 (1995): 2548–50.

———. "Liberal Democracy in India and the Dalit Critique." *Social Research: An International Quarterly* 78, no. 1 (2011): 99–122.

———. "The Indian Nation in Its Egalitarian Conception." In *Dalit Studies*, edited by Ramnarayan Rawat and S. Satyanarayana, 31–51. Durham, NC: Duke University Press, 2016.

Hari, Abhishek. "Casteism Is Rampant in Higher Education Institutions, but Is 'Wilfully Neglected': Study." *The Wire*, October 8, 2021. Accessed April 1, 2024. https://thewire.in/education/casteism-rampant-higher-education-institutions-wilfully-neglected-study.

Himanshu. "We Require Reliable Data on Caste to Meet Our Aims of Social Justice." *Mint*, April 21, 2023. Accessed April 1, 2024. https://www.livemint.com/opinion/columns/we-require-reliable-data-on-caste-to-meet-our-aims-of-social-justice-11682015797935.html.

Hindustan Times. "Call on Caste Census by Cabinet." August 19, 2010.

———. "Classifying Caste Census Data Could Be Panagariya's Toughest Job." July 18, 2015. Accessed April 1, 2024. https://www.hindustantimes.com/

india/classifying-caste-census-data-could-be-panagariya-s-toughest-job/story-SYNZijYZOYQwyNyOPjxikN.html.

———. "Pranab Justifies Caste-Based Census." May 8, 2010. Accessed April 1, 2024. https://www.hindustantimes.com/india/pranab-justifies-caste-based-census/story-k1P2KTd4hQCpyrthxASk5L.html.

Hirway, Indira. "Identification of BPL Households for Poverty Alleviation Programmes." *Economic and Political Weekly* 38, no. 45 (2003): 4803–08.

HT Correspondent. "Civil Society Movement Opposes Caste Census." *Hindustan Times*, June 11, 2010. Accessed April 1, 2024. https://www.hindustantimes.com/delhi/civil-society-movement-opposes-caste-census/story-Ef8wduE666VohYcxGazCeJ.html.

———. "Write Meri Jaati Hindustani in 2011 Caste Based Census." *Hindustan Times*. May 16, 2010. Accessed April 1, 2024. https://www.hindustantimes.com/delhi/write-meri-jaati-hindustani-in-2011-caste-based-census/story-8BlnZhh7dGAyA1IFi97FGI.html.

Indian Express. "Cabinet Meeting: Panagariya to Head Panel to 'Classify' Caste Census Data." July 17, 2015. Accessed April 1, 2024. https://indianexpress.com/article/india/india-others/cabinet-meeting-panagariya-to-head-panel-to-classify-caste-census-data/.

———. "Sonia As NAC Head Is Pseudo-Constitutional Power Center: BJP." March 30, 2010.

Irschick, Eugene. *Politics and Social Conflict in South India*. Berkeley and Los Angeles: University of California Press, 1969.

Itzigsohn, José, and Karida Brown. *The Sociology of W. E. B. Du Bois: Racialized Modernity and the Global Color Line*. New York: New York University Press, 2020.

Jaffrelot, Christophe, ed. *Hindu Nationalism: A Reader*. Delhi: Permanent Black, 2007.

———. *India's Silent Revolution: The Rise of the Lower Castes in North India*. New York: Columbia University Press, 2003.

———. "Introduction." In *The Sangh Parivar*, edited by Christophe Jaffrelot, 1–22. New Delhi: Oxford University Press, 2005.

———. "Introduction." In *Hindu Nationalism: A Reader*, edited by Christophe Jaffrelot, 3–25. Delhi: Permanent Black, 2007.

Jaffrelot, Christophe, and Sanjay Kumar. *Rise of the Plebeians? The Changing Face of the Indian Legislative Assemblies*. Delhi: Routledge, 2012.

Jain, Sachin Kumar. "Identification of the Poor: Flaws in Government Surveys." *Economic and Political Weekly* 39, no. 46 (November 20, 2004): 4981–84.

Jalan, Jyotsna, and Rinku Murgai. "An Effective 'Targeting Shortcut'? An Assessment of the 2002 Below-Poverty Line Census Method." World Bank, mimeo, 2007.

Japhet, S. and Chandan Gowda, eds. *Caste Census towards an Inclusive India*. Bangalore: Centre for the Study of Social Exclusion and Inclusive Policy, National Law School of India University, 2010.

Jasanoff, Sheila. "(No) Accounting for Expertise." *Science and Public Policy* 30, no. 3 (2003): 157–62.

———. "Science, Politics, and the Renegotiation of Expertise at EPA." *Osiris* 7 (1992): 194–217.

———. *States of Knowledge: The Co-Production of Science and the Social Order*. London and New York: Routledge, 2004.

———. *The Fifth Branch*. Cambridge, MA: Harvard University Press, 1993.

Jeffrey, Robin. "Missing from the Indian Newsroom." *The Hindu*, April 9, 2012.

Jodhka, Surinder. *Caste in Contemporary India*. New Delhi: Routledge, 2016.

Jodhka, Surinder, and Katherine Newman. "In the Name of Globalisation: Meritocracy, Productivity and the Hidden Language of Caste." *Economic and Political Weekly* 42, no. 41 (October 13, 2007): 4125–32.

Jones, Kenneth W. "Religious Identity and the Indian Census." In *The Census in British India*, edited by Norman Gerald Barrier, 73–101. New Delhi: Manohar, 1981.

Juergenseyer, Mark. *Religion as Social Vision*. Berkeley: University of California Press, 1982.

Kalaiyarasan, A., and M. Vijayabaskar. *The Dravidian Model: Interpreting the Political Economy of Tamil Nadu*. Cambridge: Cambridge University Press, 2021.

Kannabiran, Vasanth, and Kalpana Kannabiran. "Caste and Gender: Understanding Dynamics of Power and Violence." *Economic and Political Weekly* 26, no. 37 (1991): 2130–34.

Kertzer, David I., and Dominique Arel. "Censuses, Identity Formation, and the Struggle for Political Power." In *Census and Identity: The Politics of Race, Ethnicity, and Language in National Censuses*, edited by David I. Kertzer and Dominique Arel, 1–42. Cambridge: Cambridge University Press, 2002.

Kothari, Rajni, ed. *Caste in Indian Politics*. Delhi: Orient BlackSwan, 1995.

Krishnan, P. S. "Census Opportunity." *Frontline*, October 10, 2018. Accessed April 1, 2024. https://frontline.thehindu.com/the-nation/article25167014.ece.

Kumar, Arvind. "Exclusion of Pasmanda Muslims and Dalit Christians from the Scheduled Caste Quota" *South Asia Research* 43, no. 2 (2023): 192–209.

Kumar, Vijay. "Demand for Caste-Wise Census in 2011." *The Hindu*, December 24, 2009.

Kureel, P. "Indian Media and Caste." *Caste: A Journal on Social Exclusion* 2, no. 1 (2021): 97–108.

Latour, Bruno. *Science in Action*. Cambridge, MA: Harvard University Press, 1987.

Lee, Caroline W., Michael McQuarrie, and Edward T. Walker, eds. *Democratizing Inequalities: Dilemmas of the New Public Participation*. New York University Press, 2015.

Lee, Joel. *Deceptive Majority: Dalits, Hinduism, and Underground Religion*. Cambridge: Cambridge University Press, 2021.

Leonard, Dickens. "Caste-Less Tamils and Early Print Public Sphere: Remembering Iyothee Thass (1845–914)." *South Asia Research* 41, no. 3 (2021): 349–68.

"Lok Sabha (No. 14) Debate." New Delhi: Lok Sabha Secretariat, October 3, 2008. Accessed April 1, 2024. https://eparlib.nic.in/bitstream/123456789/785690/1/lsd_14_14_23-10-2008.pdf#search=null%20[2000%20TO%202009]%20 2008.

"Lok Sabha (No. 15) Debate." New Delhi: Lok Sabha Secretariat, May 6, 2010. Accessed April 1, 2024. https://eparlib.nic.in/bitstream/123456789/2965710/1/lsd_eng_15_04_06-05-2010.pdf.

"Lok Sabha (No. 15) Debate." New Delhi: Lok Sabha Secretariat, May 7, 2010. Accessed April 1, 2024. https://eparlib.nic.in/bitstream/123456789/2965711/1/lsd_eng_15_04_07-05-2010.pdf.

Louis, Prakash. "Dalit Christians: Betrayed by State and Church." *Economic and Political Weekly* 42, no. 16 (2007): 1410–14.

Loveman, Mara. *National Colors: Racial Classification and the State in Latin America*. Oxford: Oxford University Press, 2014.

———. "The Modern State and the Primitive Accumulation of Symbolic Power." *American Journal of Sociology* 110, no. 6 (2005): 1651–83.

———. "The U.S. Census and the Contested Rules of Racial Classification in Early Twentieth-Century Puerto Rico." *Caribbean Studies* 35, no. 2 (2007): 79–114.

Loveman, Mara, and Jerônimo O. Muniz. "How Puerto Rico Became White: Boundary Dynamics and Intercensus Racial Reclassification." *American Sociological Review* 72, no. 6 (2007): 915–39.

Magubane, Bernard. *The Making of a Racist State: British Imperialism and the Union of South Africa, 1875–1910*. Trenton, NJ: Africa World Press, 1996.

Maken, Ajay. "Document No. H-1102/8/2010-CD(CEN)." Ministry of State for Home Affairs, February 20, 2010.

Mandal, Dilip. "Interview by Gurinder Singh Azad, Anup Vimal, Noopur, and Anoop Kuman." *Dalit News from Kerala*, September 21, 2010. Accessed April 1, 2024. https://dalitskerala.wordpress.com/2010/09/21/.

———. "Upper-Caste Domination in India's Mainstream Media and Its Extension in Digital Media." *Economic and Political Weekly* 55, no. 46 (November 21, 2020): 34–39.

Marten, J. T. *Census of India 1921, India, Vol. 1, Part II: Tables*. Calcutta: Superintendent Government Printing, 1923.

Mehdudia, Sujay, and Siddharth Varadarajan. "Government Not for Caste Census." *The Hindu*, May 5, 2010. Accessed April 1, 2024. https://www.thehindu.com/news/national/Government-not-for-caste-census/article16298373.ece.

Mehrotra, Santosh, and Harsh Mander. "How to Identify the Poor? A Proposal." *Economic and Political Weekly* 44, no. 19 (May 9, 2009): 37–44.

Mehta, Pratap Bhanu. "My Caste and I." *Indian Express*, May 12, 2010. Accessed April 1, 2024. https://indianexpress.com/article/opinion/columns/my-caste-and-i/.

Ministry Home Affairs, P. K. Banerjee, ORGI. "State-Wise Number of EBs/ Sub-Blocs for Use in the SECC." No. 32/4/20-CD (CEN), November 8, 2011.

Ministry of Housing and Urban Poverty Alleviation, Government of India. "Instruction Manual for Enumerators, Socio-Economic and Caste Census 2011-Urban." Government of India. Accessed March 25, 2024. Accessed April 1, 2024. https://mohua.gov.in/upload/uploadfiles/files/31375(1).pdf.

Ministry of Personnel, Public Grievances and Pensions, Government of India. *Annual Report 2013–2014*. 2014.

———. *Annual Report 2016–2017*. 2017. Accessed April 1, 2024. https://dopt.gov.in/sites/default/files/AnnualReport_2016-17_ENG.pdf.

———. "Brochure on Reservation for SCs, STs & OBCs in Services." 2014.

Ministry of Rural Development. B. K. Sinha. "Socio Economic Caste Census Guidelines (with Attachments)." No. Q-14015/04/2011 AI (RD), June 24, 2011. Accessed April 1, 2024. https://rural.gov.in/sites/default/files/Guidlines_27062011_0.pdf.

———. "Socio Economic Caste Census Procedure." File D.R. No. Q-1615/04/2011AR(RD), June 24, 2011.

———. "Socio Economic Census to Identify Below Poverty Line (BPL) Families." No. Q-14016/6/2011 AI-(RD), May 30, 2011.

Ministry of Rural Development, N. K. Shau. "Letter to Primary Secretary of All State Governments and Nodal Officers, RGI, HUPA, and CPSUs." Doc. No. A-16015/07/2011-AI (RD), December 2, 2011.

Ministry of Rural Development, P. K. Padhy. "Instructions for Opening a Dedicated Account for the Socio-Economic Survey-2011." Doc. No. A 16025/8/2006-AI (RD), July 1, 2011. Accessed April 1, 2024. https://rural.gov.in/sites/default/files/Guidelines%20Norms%20for%20Utilization%20of%20Funds%20for%20conducting%20SECC%202011_1.pdf.

Ministry of Rural Development, Sanjay Kumar Rakesh. "Infrastructure Requirements for Charge Centres Data Center (Tehsil Level) for SECC 2011." Do. No. 14016/6/2011AI (RD), July 11, 2011. Accessed April 1, 2024.

https://rural.gov.in/sites/default/files/Letter%20dated%2011.7.2011%20 Regarding%20Infrastructure%20requirements%20for%20Charge%20 Centres_0.pdf.

Ministry of Rural Development and Ministry of Housing and Urban Poverty Alleviation, Government of India. "Supervisor's Manual, Socio-Economic and Caste Census." Government of India, June 20, 2012. Accessed April 1, 2024. https://rural.gov.in/sites/default/files/Supervisory%20Manual%2020% 20June%202012%20%28English%29_1.pdf.

Ministry of Rural Development, Fifteenth Lok Sabha Committee on Estimates. *National Social Assistance Program, 4th Report (2014–2015)*. Delhi: Lok Sabha Secretariat, April 29, 2015.

Ministry of Rural Development, Government of India. "Caste Tribe Status of Households (Rural)." Accessed July 14, 2023. https://secc.gov.in/ getCasteProfileRuralNationalReport.htm.

———. "Guidelines to States for Conducting Pilot Socio-Economic Survey 2010," July 2010. Accessed April 1, 2024. https://rural.gov.in/sites/default/ files/Guidelines_4_States.pdf.

———. "Instruction Manual for Enumerators, Socio-Economic and Caste Census 2011-Rural." Government of India. Accessed March 25, 2024. https://rural.gov.in/sites/default/files/Instruction%20Manual%20for%20 Enumerators%28English%29_1.pdf.

———. "SEC Census Homepage." Accessed July 14, 2023. https://secc.gov.in/ homepage.htm.

———. "Socio Economic & Caste Census 2011 in Rural India." Government of India, July 25, 2011. Accessed April 1, 2024. https://rural.gov.in/sites/default/ files/Caste%20Census%20book%2017-11-2011_0.pdf.

———. "Socio Economic & Caste Census 2011 in Rural India." Accessed September 30, 2023. Accessed April 1, 2024. https://secc.gov.in/ homePageUrban.htm.

———. "Socio Economic and Caste Census 2011 Questionnaire Rural." July 2011. Accessed April 1, 2024. https://rural.gov.in/sites/default/files/ Questionnaire%20for%20Socio%20Economic%20and%20Caste%20 Census-2011_1.pdf.

———. "Socio-Economic Survey 2010: Supervisor Manual." Government of India, 2010.

Misra, Dipak. *Census Commissioner and Others v. R. Krishnamurthy*, No. Civil Appeal No. 9996. (Supreme Court of India November 7, 2014).

Mitchell, Timothy. *Rule of Experts: Egypt, Techno-Politics, Modernity*. Berkeley: University of California Press, 2002.

Mohan, P. S. *Modernity of Slavery*. New Delhi: Oxford University Press, 2015.

Mondol, Seik Rahim. "Social Stratification, OBCs and Muslims." *Economic and Political Weekly* 38, no. 46 (November 15, 2003): 4892–97.

"Movement against Inclusion of Caste in Census Gains Impetus." *The Hindu*, June 12, 2010.

Mukherjee, Neela. "Political Corruption in India's Below the Poverty Line (BPL) Exercise: Grassroots' Perspectives on BPL in Perpetuating Poverty and Social Exclusion." Development Tracks in Research, Training and Consultancy. New Delhi, 2005.

Nanda, Amulya Ratna. "Census of India 1991: Union Primary Census Abstract for SCs and STs, Paper 1." Registrar General and Census Commissioner, India, 1993.

Natarajan, Dandapani. *Indian Census through a Hundred Years, Census of India 1971*. Delhi: Office of the Registrar General and Census Commissioner, India, 1972.

National Commission for Scheduled Tribes. "NCST Handbook." Government of India, 2015.

Nayak, Rajesh Kumar. "Quest for Equality and Identity: The 'Pasmanda' Muslim Discourse in Post-1947 Bihar." *Proceedings of the Indian History Congress* 74 (2013): 961–71.

Nehru, Jawaharlal. *The Discovery of India*. New York: Anchor Books, 1960.

Nobles, Melissa. *Shades of Citizenship: Race and the Census in Modern Politics*. Stanford, CA: Stanford University Press, 2000.

Office of the Registrar General and Census Commissioner, India, Ministry of Home Affairs. "Census of India 2011: Household Schedule." Government of India, 2011.

———. "Know Your Enumerator," February 10, 2020. Accessed April 1, 2024. http://censusindia.gov.in/2011-common/enumeratorprofile.html.

———. "Notification S.O.120(E) [Dated: 1/7/2020], [F. No.9/7/2019-CD(CEN)3]." *Gazette of India*, January 9, 2020.

———. "REGD. Notification D. L.-33004/99." *Gazette of India*, March 11, 2022.

Omvedt, Gail. *Buddhism in India: Challenging Brahmanism and Caste*. New Delhi: Sage, 2003.

———. *Cultural Revolt in a Colonial Society*. New Delhi: Manohar, 2023.

———. *Dalits and the Democratic Revolution*. Delhi: Sage, 1994.

Paik, Shailaja. "Amchya Jalmachi Chittarkatha (The Bioscope of Our Lives): Who Is My Ally?" *Economic and Political Weekly* 44, no. 40 (October 3, 2009): 39–47.

———. *Dalit Women's Education in Modern India: Double Discrimination*. London and New York: Routledge, 2014.

———. *The Vulgarity of Caste: Dalits, Sexuality, and Humanity in Modern India*. Stanford, CA: Stanford University Press, 2022.

Paliwal, Ankur. "How India's Caste System Limits Diversity in Science—in Six Charts." *Nature* 613, no. 7943 (2023): 230–34.

Panda, Sitakanta. "Political Connections and Elite Capture in a Poverty Alleviation Programme in India." *The Journal of Development Studies* 51, no. 1 (2015): 50–65.

Parsai, Gargi. "It's for States to Identify BPL Families." *The Hindu*, January 14, 2011.

Paschel, Tianna S. *Becoming Black Political Subjects*. Princeton: Princeton University Press, 2016.

Phule, Jotirao. "Cultivator's Whipcord (1883)." In *Selected Writings of Jotirao Phule*, edited by G. P. Deshpande, 113–189. New Delhi: Leftworld Books, 2002.

———. "*From* The Book of the True Faith (1891)." In *Selected Writings of Jotirao Phule*, edited by G. P. Deshpande, 224–36. New Delhi: Leftworld Books, 2002.

———. "Opinion from Jotteerao Govindrao Phulay on Note No. II, by Mr. B. M. Malabari on Enforced Widowhood." In *Selected Writings of Jotirao Phule*, edited by G.P. Deshpande, 195–197. New Delhi: Leftworld Books, 2002.

———. *Selected Writings of Jotirao Phule*. Edited by G. P. Deshpande. New Delhi: LeftWord Books, 2002.

———. "Slavery (1873)." In *Selected Writings of Jotirao Phule*, edited by G. P. Deshpande, 22–99. New Delhi: Leftworld Books, 2002.

Plowden, W. C. *Report on the Census of British India Taken on the 17th February 1881*. London: Eyre and Spottiswoode, 1883.

"Population of SCs/STs/OBCs, Lok Sabha Question No. 3925." New Delhi: Ministry of Home Affairs, February 18, 2014. Accessed April 1, 2024. https://www.mha.gov.in/MHA1/Par2017/pdfs/par2014-pdfs/ls-180214/3925.pdf.

Porter, Theodore M. *The Rise of Statistical Thinking, 1820–1900*. Princeton: Princeton University Press, 1986.

Prahbu, Sunil. "Pranab Says Current Census Will Include Caste, after All." NDTV, May 7, 2010. Accessed April 1, 2024. https://www.ndtv.com/india-news/pranab-says-current-census-will-include-caste-after-all-417372.

Prakash, G. *Bonded Histories*. Cambridge: Cambridge University Press, 1990.

Purohit, Kunal. "How a Landmark Caste Census in India Threatens Modi's Grip on Power." *Al Jazeera*, October 4, 2023. Accessed April 1, 2024. https://www.aljazeera.com/features/2023/10/4/how-a-landmark-caste-census-in-india-threatens-modis-grip-on-power.

Radhakrishnan, P. "Caste Factor in Census 2011 Will Pollute India's Governance." *DNA*, July 12, 2011.

"Rajya Sabha Debate, Session 218." New Delhi: Rajya Sabha Secretariat, December 9, 2009. Accessed April 1, 2024. https://cms.rajyasabha.nic.in/UploadedFiles/Debates/OfficialDebatesDatewise/Floor/218/F09.12.2009.pdf.

"Rajya Sabha Debate, Session 218." New Delhi: Rajya Sabha Secretariat, December 18, 2009. Accessed April 1, 2024. https://cms.rajyasabha.nic.in/UploadedFiles/Debates/OfficialDebatesDatewise/Floor/218/F18.12.2009.pdf.

"Rajya Sabha Question Number 2698." New Delhi: Ministry of Home Affairs, August 10, 2016.

"Rajya Sabha Unstarred Question No. 2686 by Dr. John Brittas, Answered by Ministry of Personnel, Public Grievances and Pensions," March 23, 2023. Accessed April 1, 2024. https://sansad.in/getFile/annex/259/AU2686.pdf?source=pqars.

Ram, F., S. K. Mohanty, and Usha Ram. "Understanding the Distribution of BPL Cards." *Economic and Political Weekly* 44, no. 7 (2009): 67–71.

Ramaiah, A. "Identifying Other Backward Classes." *Economic and Political Weekly* 27, no. 23 (June 6, 1992): 1203–07.

Rege, Sharmila. *Against the Madness of Manu: B. R. Ambedkar's Writings on Brahmanical Patriarchy*. Delhi: Navayana, 2013.

———. *Writing Caste/Writing Gender: Reading Dalit Women's Testimonies*. Delhi: Zubaan, 2006.

Risley, H. H., and E. A. Gait. *Census of India 1901, Vol 1A India, Part II. Tables*. Calcutta: Office of Superintendent of Government Printing, India, 1903.

Roberts, Dorothy. "Sources of Commitment to Social Justice." *Roger Williams University Law Review* 4, no. 1 (1998): 175–203.

Rodríguez-Muñiz, Michael. "Cultivating Consent: Nonstate Leaders and the Orchestration of State Legibility." *American Journal of Sociology* 123, no. 2 (September 2017): 385–425. https://doi.org/10.1086/693045.

Roy, Anirban Guhaand, and Subhash Pathak. "Decoding Findings of Bihar's Caste Survey Report." *Hindustan Times*, November 8, 2023. Accessed April 1, 2024. https://www.hindustantimes.com/india-news/decoding-findings-of-bihars-caste-survey-report-101699396896631.html.

Roy, Indrajit. "'New' Lists for 'Old': (Re-)Constructing the Poor in the BPL Census." *Economic and Political Weekly* 46, no. 22 (May 28, 2011): 82–91.

Sachar, Rajindar, Saiyid Hamid, T. K. Oommen, M. A. Basith, Rakesh Basant, Akhtar Majeed, and Abusaleh Shariff. "Social, Economic and Educational Status of the Muslim Community of India." Prime Minister's High Level Committee. New Delhi: Cabinet Secretariat, November 2006. Accessed April 1, 2024. https://www.education.gov.in/sites/upload_files/mhrd/files/sachar_comm.pdf.

Samarendra, Padmanabh. "Census in Colonial India and the Birth of Caste." *Economic and Political Weekly* 46, no. 33 (2011): 51–58.

Saperstein, Aliya, and Andrew Penner. "Beyond the Looking Glass: Exploring Fluidity in Racial Self-Identification and Interviewer Classification." *Sociological Perspectives* 57, no. 2 (2014): 186–207.

———. "Racial Fluidity and Inequality in the United States." *American Journal of Sociology* 118, no. 3 (2012): 676–727.

Savarkar, Vinayak Damodar. "Hindutva: Who Is a Hindu." In *Hindu Nationalism: A Reader*, edited by Christophe Jaffrelot, 87–96. Delhi: Permanent Black, 2007.

Scott, James. *Seeing Like a State*. New Haven: Yale University Press, 1998.

Shanmugavelan, Murali. "May the Myth of Castelessness Die." *LOGIC(S)*, no. 19 (2023). Accessed April 1, 2024. https://logicmag.io/supa-dupa-skies/may-the-myth-of-castelessness-die/.

Shepherd, Kancha Illaiah. *God as Political Philosopher: Buddha's Challenge to Brahmanism*. 2nd edition. New Delhi: Sage, 2019.

———. *Why I Am Not a Hindu: A Sudra Critique of Hindutva Philosophy, Culture and Political Economy*. 2nd edition. New Delhi: Sage, 2019.

Shrikant, L. M. *Report of the Commissioner of Scheduled Castes and Scheduled Tribes*. New Delhi: Manager, Government of India Press, 1956.

Singh, Nirmal. *Dalits in Indian Democracy*. Delhi: Kalpaz Publications, 2018.

Singh, Pritam. "Numbers as a Means to Power: Politics of Caste as a Census Category in Colonial India c 1871–1941." *Economic and Political Weekly* 57, no. 18 (April 30, 2022): 63–67.

Singh, Vivek Kumar. "Counting Caste: A Step toward Radical Anticasteism." *Prabuddha: Journal of Social Equality* 7, no. 1 (2022): 1–13.

Soundararajan, Thenmozhi. *The Trauma of Caste: A Dalit Feminist Meditation on Survivorship, Healing, and Abolition*. Huichin, unceded Ohlone land, aka Berkeley, CA: North Atlantic Books, 2022.

Srinivas, M. N. *Caste in Modern India and Other Essays*. London: Asia Publishing House, 1962.

Srinivas, M. N. "An Obituary on Caste as a System." *Economic and Political Weekly* 38, no. 5 (February 1, 2003): 455–59.

Standing Committee on Rural Development. *Twenty Seventh Report*. Delhi: Lok Sabha Secretariat, August 31, 2016.

Starr, Kelsey Jo, and Neha Sahgal. "Measuring Caste Data in India." *Pew Research Center Decoded* (blog), June 21, 2021. Accessed April 1, 2024. https://www.pewresearch.org/decoded/2021/06/29/measuring-caste-in-india/.

Subrahmanyam, R. *State of Maharashtra v. Union of India*, "Reply Affidavit on the Behalf of Respondent-UOI," No. Write Petition (Civil) no. 841 of 2021 (Supreme Court of India September 21, 2021).

Subramanian, Ajantha. "Making Merit: The Indian Institutes of Technology and the Social Life of Caste." *Comparative Studies in Society and History* 57, no. 2 (2015): 291–32.

———. *The Caste of Merit*. Cambridge, MA: Harvard University Press, 2019.

Sundaram, K. "On Identification of Households Below Poverty Line in BPL Census 2002: Some Comments on the Proposed Methodology." *Economic and Political Weekly* 38, no. 9 (March 1, 2003): 896–901.

Swaminathan, Madhura. "Public Distribution System and Social Exclusion." *The Hindu*, May 7, 2008.

Tewari, Ruhi, and Vijaita Singh. "Rural Development to Home to Social Justice, No One Wants to Own Undisclosed Caste Data," *Indian Express*, July 8, 2015. Accessed April 1, 2024. https://indianexpress.com/article/india/india-others/rural-development-to-home-to-social-justice-no-one-wants-to-own-undisclosed-caste-data/.

Tharamangalam, J. "Caste among Christians in India." In *Caste: Its Twentieth Century Avatar*, edited by M. N. Srinivas, 263–91. Delhi: Penguin Random House India, 1997.

The Hindu. "Census 2011 to Include Caste." May 8, 2010.

———. "Court Directs Government to Conduct Caste-Wise Census." May 13, 2010.

———. "Issue of Caste-Wise Census Taken up with Centre." January 24, 2010.

———. "Movement against Caste-Based Census Launched." June 11, 2010. Accessed April 1, 2024. https://www.thehindu.com/news/cities/Delhi/Movement-against-caste-based-Census-launched/article16242760.ece.

———. "Plea for Caste-Wise OBC Enumeration." October 16, 2009.

———. "Ramadoss Seeks Caste-Wise Census." September 19, 2009.

The Wire. "Bihar Passes Bills Raising Caste-Based Quota from 50% to 65%, Taking Total Reservation to 75%." November 9, 2023. Accessed April 1, 2024. https://thewire.in/government/bihar-extends-caste-based-reservation-from-50-percent-to-65-percent.

———. "No Caste Enumeration Other than SC and ST in Upcoming Census, Says Govt." July 20, 2021. Accessed April 1, 2024. https://thewire.in/government/no-caste-enumeration-other-than-sc-and-st-in-upcoming-census-says-govt.

———. "Patna High Court Dismisses Petitions against Bihar Caste Census." August 1, 2023. Accessed April 1, 2024. https://thewire.in/law/patna-high-court-dismisses-petitions-against-bihar-caste-census.

Thomas, Abraham. "Pleas against Caste Census in Bihar Merely for Publicity: Supreme Court." *Hindustan Times*, January 21, 2023. Accessed April 1, 2024. https://www.hindustantimes.com/india-news/sc-rejects-pleas-against-bihar-caste-census-calls-them-publicity-litigation-101674213641883.html.

Thomas, Sonja. *Privileged Minorities: Syrian Christianity, Gender, and Minority Rights in Postcolonial India.* Seattle: University of Washington Press, 2018.

Thompson, Debra. *The Schematic State: Race, Transnationalism, and the Politics of the Census.* Cambridge: Cambridge University Press, 2016.

Thorat, Sukhadeo. *Dalits in India: Search for a Common Destiny.* Delhi: Sage, 2009.

Times of India. "Caste Census Should Have Continued after Independence: Pranab." May 8, 2010.

———. "GoM Divided over Caste Census." July 1, 2010.

Tripathi, Rahul. "OBC Count to Be Part of Census 2021, 3 Decades after Mandal Commission." *Indian Express*, September 1, 2018. Accessed April 1, 2024. https://indianexpress.com/article/india/obc-count-to-be-part-of-census-2021-3-decades-after-mandal-5334643.

Trivedi, Prashant K., Srinivas Goli, Fahimuddin, and Surinder Kumar. "Does Untouchability Exist among Muslims? Evidence from Uttar Pradesh." *Economic and Political Weekly* 51, no. 15 (April 9, 2016): 32–36.

Upadhyay, Sparsh. "Patna High Court Reserves Verdict in Pleas Challenging Bihar Govt's Decision to Conduct Caste-Based Survey in State." *Live Law*, July 7, 2023. Accessed April 1, 2024. https://www.livelaw.in/high-court/patna-high-court/patna-high-court-reserves-verdict-pleas-challenging-bihar-government-decision-conduct-caste-based-survey-232205.

Vaghela, Palashi, Steven J. Jackson, and Phoebe Sengers. "Interrupting Merit, Subverting Legibility: Navigating Caste in 'Casteless' Worlds of Computing." In *Proceedings of the 2022 CHI Conference on Human Factors in Computing Systems (CHI '22)*, Article 545:1–20. New York, NY: Association for Computing Machinery, 2022. https://doi.org/10.1145/3491102.3502059.

Venkatesan, J. "Apex Court Moved for OBC Census." *The Hindu*, April 15, 2010.

———. "Moily Defends Caste-Based Census." *The Hindu*, May 21, 2010. Accessed April 1, 2024. https://www.thehindu.com/news/national/Moily-defends-caste-based-census/article16302260.ece.

———. "Supreme Court Declines to Entertain PMK's Petition." *The Hindu*, April 10, 2009.

Verma, V. S. "Census of India 1981: A Handbook of Population Statistics." Registrar General and Census Commissioner, India, 1988.

Vij, Shivani, and Shrutanjaya Bhardwaj. "Can Bihar Increase Its Reservation Pool? | Explained." *The Hindu*, December 11, 2023.

Vijaya Karnataka. "Urban Caste Census." September 16, 2011.

Vijayanunni, M. "BPL Surveys Are Mere BPL Surveys, Not Caste Census: Dr M Vijayanunni." *Economic Times*, July 31, 2011. Accessed April 1, 2024. https://economictimes.indiatimes.com/opinion/interviews/bpl-surveys-are-mere-bpl-surveys-not-caste-census-dr-m-vijayanunni/articleshow/9423430.cms.

———. "Caste Can't Be Caste Off." *Indian Express*, November 19, 1999.

———. "Counting Caste: Methodology." In *Caste Census towards an Inclusive India*, edited by Japhet, S. and Gowda, Chandan, 39–50. Bangalore: Centre for the Study of Social Exclusion and Inclusive Policy, National Law School of India University, 2010.

Vijayanunni, M., M. N. Rao, Sukhadeo Thorat, Satish Deshpande, Yogendra Yadav, S. Japhet, Chandan Gowda, Valerian Rodrigues, and Ravi Varma Kumar. "Letter to the Group of Ministers on Caste Census." *The Hindu*, August 13, 2010.

Viswanath, Rupa. *The Pariah Problem: Caste, Religion, and the Social in Modern India*. New York, NY: Columbia University Press, 2014.

Vithayathil, Trina. "Counting Caste: Census, Politics and Castelessness in India." *Politics and Society* 46, no. 4 (2018): 455–84.

Vithayathil, Trina, Diana Graizbord, and Cedric de Leon. "The Retreat to Method: The Aftermath of Elite Concession to Civil Society in India and Mexico." *Studies in Comparative International Development* 54, no. 1 (2019): 19–39.

Vyas, Neena. "BJP Ducks Firm Response to Pranab Query on Census." *The Hindu*, August 7, 2010.

Waghmore, Suryakant. *Civility against Caste*. Delhi: Sage, 2013.

Waterfield, Henry. *Memorandum on the Census of British India of 1871–72*. London: H. M. Stationery Office, 1875.

Wilkinson, Steven Ian. "India, Consociational Theory, and Ethnic Violence." *Asian Survey* 40, no. 5 (2000): 767–91.

Witsoe, Jeffrey. *Democracy against Development: Lower-Caste Politics and Political Modernity in Postcolonial India*. Chicago: University of Chicago Press, 2013.

Yadav, Yogendra. "Reconfiguration in Indian Politics: State Assembly Elections, 1993–95." *Economic and Political Weekly* 31, no. 2/3 (January 13, 1996): 95–104.

———. "Understanding the Second Democratic Upsurge." In *Transforming India: Social and Political Dynamics of Democracy*, edited by Francine Frankel, Zoya Hasan, Rajeev Bhargava, and Balveer Arora, 89–119. Delhi: Oxford University Press, 2002.

———. "Why Caste Should Be Counted." *The Hindu*, May 14, 2010.

Yengde, Suraj. *Caste Matters*. Delhi: Penguin Random House India Private Limited, 2019.

Zainuddin, Sayyed. "Islam, Social Stratification and Empowerment of Muslim OBCs." *Economic and Political Weekly* 38, no. 46 (November 15, 2003): 4898–901.

Zelliot, Eleanor. *Ambedkar's World: The Making of Babasaheb and the Dalit Movement*. New Delhi: Navayana, 2013.

Index